高校专门用途英语（ESP）系列教材

A PRACTICAL COURSE of
EST Reading
科技英语阅读实用教程

主编：谢小苑

编者：王珺琳　徐智鑫　王秀文
　　　李　迟　钱　菁　高乃晶

清华大学出版社
北京

内 容 简 介

本书系统介绍科技英语的语言特点及科技文献的阅读技巧与方法，通过各种科技文体的阅读实践，使学生掌握常用的阅读技能并达到一定的熟练程度。

本书共 12 单元，每单元由 3 部分组成。第一部分"讲座"，主要讲解科技文献的阅读技巧和方法。第二部分"阅读"，课文内容突出中西方科技史特色，介绍人类发展进程中各个时期科学和技术的发展、科技大发明和大发现、科学家故事等。第三部分"拓展阅读"，目的是给学生提供更多的阅读实践机会；同时，通过大量阅读增加词汇量，开阔视野。

本书具有技巧与实践相结合、知识性与实用性相结合、系统性与针对性相结合的特色，适合作为高校非英语专业学生的科技英语阅读教材，也可供广大科技人员和科技英语爱好者阅读参考。

本书的练习参考答案和教学课件请在 ftp://ftp.tup.tsinghua.edu.cn/ 上进行下载。

版权所有，侵权必究。侵权举报电话：010-62782989 13701121933

图书在版编目（CIP）数据

科技英语阅读实用教程/谢小苑主编. —北京：清华大学出版社，2020.8
高校专门用途英语（ESP）系列教材
ISBN 978-7-302-55179-9

Ⅰ.①科… Ⅱ.①谢… Ⅲ.①科技学术—英语—阅读教学—高等学校—教材 Ⅳ.① N43

中国版本图书馆 CIP 数据核字（2020）第 049562 号

责任编辑：倪雅莉　刘　艳
封面设计：子　一
责任校对：王凤芝
责任印制：丛怀宇

出版发行：清华大学出版社
网　　址：http://www.tup.com.cn, http://www.wqbook.com
地　　址：北京清华大学学研大厦 A 座　　邮　编：100084
社 总 机：010-62770175　　邮　购：010-62786544
投稿与读者服务：010-62776969, c-service@tup.tsinghua.edu.cn
质量反馈：010-62772015, zhiliang@tup.tsinghua.edu.cn

印 装 者：小森印刷霸州有限公司
经　　销：全国新华书店
开　　本：170mm×230mm　　印　张：24.5　　字　数：396 千字
版　　次：2020 年 8 月第 1 版　　印　次：2020 年 8 月第 1 次印刷
定　　价：75.00 元

产品编号：086750-01

Preface 前言

随着世界经济全球化、科技一体化、文化多元化进程的加速,国家和社会对非英语专业学生的英语应用能力提出了新的要求,对大学英语教学提出了新的挑战。新出台的《大学英语教学指南》明确要求大学英语课程教学内容从单一的通用英语向"通用英语+"型,即通用英语+专门用途英语和跨文化交际的一体两翼转变(王守仁,2016)。为适应新形势并满足新时期国家和社会对人才培养的需求,南京航空航天大学加强专门用途英语类和跨文化交际类课程建设,突出知识构建、能力培养和文化素质的提高,使学生的英语知识、能力、素质得到协调发展,培养具有责任意识、创新精神、国际视野和人文情怀的社会栋梁和工程英才,实现新形势下大学英语教学的可持续发展,促进大学英语课程的长远发展。

《科技英语翻译实用教程》和《科技英语阅读实用教程》是南京航空航天大学的大学英语教学改革探索和专门用途英语(ESP)教学团队教学实践积累的成果,旨在帮助学生顺利阅读和翻译所学专业的英语文献和资料。我们期待该系列教材能为学生知识-能力-素质的协调发展和英语应用能力的提升提供一个良好的契机和新的生长点。

《科技英语阅读实用教程》具有以下特色:

1. 技巧与实践相结合

本书在编排上力求有所创新,突显技巧与实践的紧密结合。本书以"讲座"的形式讲解科技英语的语言特点、科技文献的阅读技巧与方法,以"阅读"及"拓展阅读"作为阅读实践内容,介绍由人类发展进程中各个时期的科技发展、科技大发明和大发现、科学家故事连缀成的中外科技史。阅读技巧"讲座"既体现科技文献阅读的技巧和方法,又提供大量的实例和练习,做到讲练结合;"阅读"及"拓展阅读"的阅读实践以阅读技巧和方法为指导,做到学用结合。

2. 知识性与实用性相结合

本书的编写注重实用性，内容力求做到深入浅出、通俗易懂，目的是帮助学生掌握科技文献阅读的技能与方法。本书所选课文、实例和练习内容围绕中外科技史，包括科学和技术发展介绍、科技大发明和大发现、科学家故事等。学生在学习阅读技巧与进行阅读实践的同时，也可以了解人类发展进程中科学与技术的发展轨迹，领略每一项发明创造的深远影响，感受科技发展的内在脉络以及科技发展的传承性，以提高科学文化素养。

3. 系统性与针对性相结合

本书编者认为，一本较好的科技英语教材应该研究学习者本身的特点和与之关联的工作、社会、未来等需求，应该结合学习者所学专业及相关学科。因此，本书针对高校大学生（尤其是理工科学生）的特点，系统介绍科技英语的语言特点、科技文献的阅读技巧与方法，重点讲解学生在阅读中常碰到的各种问题及解决方法，选择学生未来工作中会经常接触到的文体形式及来源于实际运用中的语言材料，帮助学生尽快了解其感兴趣的领域，以适应实际工作和满足社会的需要。

本书共 12 单元，每单元由 3 部分组成。第一部分"讲座"，主要讲解科技文献的阅读技巧与方法。第二部分"阅读"，课文内容突出中西方科技史特色，介绍人类发展进程中各个时期科学和技术的发展、科技大发明和大发现、科学家故事等。第三部分"拓展阅读"，目的是给学生提供更多的阅读实践机会；同时，通过大量阅读增加词汇量，开阔视野。

本书得以付梓，离不开方方面面的支持。首先，本书为南京航空航天大学 2019 年校级教育教学改革项目（精品教材建设专项）的研究成果，我们对学校项目的资助表示感谢。其次，感谢南京航空航天大学的各级领导，尤其是教务处及教材科、外国语学院领导的支持，他们在经费和政策上的大力支持为本书的顺利完成提供了有力保障。再次，感谢参与教材修订的 ESP 教学团队成员，他们在《科技英语阅读》（2014）的基础上，对本书内容进行补充和更新，日臻完善。本书在编写过程中，参考了国内外出版的相关书刊并引用了部分资料，在此向有关作者和单位表示诚挚的感谢。最后感谢清华大学出版社领导及编辑，他们为本书出版付出了辛勤的劳动。

　　由于编者水平和经验有限，书中欠妥与谬误之处在所难免，祈请同行专家和广大读者多多斧正，以便今后修订完善。

编者

2019 年 10 月于南京航空航天大学

Contents 目录

Unit One
Historical Development of Science and Technology 1
Part A Lecture 了解科技英语的语言特点 2
Part B Reading Technological Evolution of Humankind 12
Part C Extended Reading Invention of Wheel 21

Unit Two
Ancient Civilizations 25
Part A Lecture 提高科技英语阅读理解能力 26
Part B Reading Ancient Technology 43
Part C Extended Reading Development of Civilization 51

Unit Three
History of Science in Early Cultures 55
Part A Lecture 掌握科技英语文献的阅读方法 56
Part B Reading Science in the Greek World 76
Part C Extended Reading Plato and Aristotle 83

Unit Four
History of Science and Technology in Ancient China 87

Part A Lecture 确定中心思想 88
Part B Reading Science and Technology of the Tang Dynasty 109
Part C Extended Reading Science in Ancient China 119

Unit Five
Science and Technology in the Middle Ages and Renaissance 125

Part A Lecture 辨认事实与细节 126
Part B Reading Medieval Science and Technology 141
Part C Extended Reading Johannes Gutenberg and His Invention of Movable Type Printing 150

Unit Six
Impact of Science in Europe 155

Part A Lecture 识别逻辑衔接词 156
Part B Reading New Ideas During the Scientific Revolution 166
Part C Extended Reading Science in the Age of Enlightenment 180

Unit Seven
Industrial Revolution 185

Part A Lecture 识别段落类型 186
Part B Reading The Steam Engine—A New Source of Power 202
Part C Extended Reading Machine Tools 208

Unit Eight
Modern Science and Technology (I) 213

- **Part A Lecture** 了解衔接与连贯手段的使用 214
- **Part B Reading** Science and Technology in the Electrical Age 228
- **Part C Extended Reading** Development of Natural Sciences 236

Unit Nine
Modern Science and Technology (II) 247

- **Part A Lecture** 猜测词义 248
- **Part B Reading** Chips and the Internet 263
- **Part C Extended Reading** In 1900, They Never Imagined 271

Unit Ten
Modern Development in China 283

- **Part A Lecture** 推理 284
- **Part B Reading** Modern Development in China 306
- **Part C Extended Reading** Yuan Longping—Father of Hybrid Rice 312

Unit Eleven
Scientific Discoveries and Inventions 317

- **Part A Lecture** 结论 318
- **Part B Reading** Technology's Ubiquitous Reach 333
- **Part C Extended Reading** The Telephone and Its Inventor 340

Unit Twelve
Great Men and Women 345
Part A Lecture 判定作者的观点、态度和语气 346
Part B Reading Three Great Physicists 365
Part C Extended Reading Madame Curie and Radium 374

References 377

Unit One

Historical Development of Science and Technology

Part A Lecture

了解科技英语的语言特点

随着国际间科技、经济和文化交往的日益扩大，专门用途英语（English for Specific Purposes，简称ESP）应运而生。专门用途英语是指与某种特定职业或学科相关的英语，是根据学习者特定目的和需要而开设的英语课程，如商务英语、医学英语、科技英语等，其目的是培养学生在一定工作环境中运用英语开展工作的交际能力。科技英语（English for Science and Technology，简称EST）是专门用途英语中最重要的一个分支。随着世界科技发展的日新月异和我国对外科技交流的日益频繁，科技英语的重要性越来越清楚地显现出来。因此，为了更好地促进科技交流，促进国民经济的发展，同时也为了提高自身的专业素质，保证职业生涯的成功，理工学科背景的大学生有必要学习科技英语课程。本单元将分析科技英语及其语言特点。

1 科技英语

科技英语是一种重要的英语语体，是随着科学技术的发展而形成的一种独立的文体形式。科技英语泛指与科学技术有关的书面或口头的英语文献，既涵盖自然科学领域的各种知识和技术，也包括社会科学的各个领域，如：用英语撰写的有关自然科学和社会科学的情报资料、学术著作、论文、实验报告、产品和专利的说明书等。为了准确、简洁、明了地叙述自然现象和事实及其发展过程、性质和特征，科技人员常常使用一些典型的句型和大量的专业术语，因而形成科技英语自身的特色。试分析例1，找出科技英语特征。

Unit One Historical Development of Science and Technology

It is nothing else than impurities prenatally inherent in ore that seriously affect the quality of the latter, which is formed as a result of geological vicissitudes including diastrophic movement, eruption of volcano, sedimentation, glaciation and weathering etc., under the action of which pyrogenic rocks etc., come into being, some of which exist in a stage of symbiosis, the main cause of the absence of pure rocks in nature, wherein lies the reason for the need of separation technology and apparatus, namely, ore-dressing devices and equipment, (which has been) so far impotent to meet the requirements of metallurgical industry (which) the scientists make every endeavor to elevate to a new high by laser separation. (许建平, 2003)

分析：从例1可以看出，该段具有明显的科技英语特征：1）逻辑严密，结构紧凑，陈述客观，表达清晰；2）专业词汇多，如vicissitudes（变迁）、diastrophic（地壳变迁）、sedimentation（沉积）、glaciations（冰川）、weathering（风化）、pyrogenic rocks（火成岩）、symbiosis（共生现象）等；3）从句多，如该句主句为 It is...in ore; 从句 that seriously affect...by laser separation; 主语从句中包含的定语从句有：which is formed..., under the action of which... come into being, wherein (=in which), (which have been) so far impotent to..., metallurgical industry (which) the scientists make every endeavor to...。

根据主从结构和从句之间的关系，这个复杂长句可分成六个小句：

① It is nothing else than impurities prenatally inherent in ore that seriously affect the quality of the latter,

② which is formed as a result of geological vicissitudes including diastrophic movement, eruption of volcano, sedimentation, glaciation and weathering etc.,

③ under the action of which pyrogenic rocks etc., come into being,

④ some of which exist in a stage of symbiosis, the main cause of the absence of pure rocks in nature,

⑤ wherein lies the reason for the need of separation technology and apparatus, namely, ore-dressing devices and equipment,

⑥ (which has been) so far impotent to meet the requirements of metallurgical industry the scientists make every endeavor to elevate to a new high by laser separation.

  根据上下文和全句逻辑关系，本句可以这样理解：影响矿石质量的不是别的东西，而是矿石中天然固有的杂质。矿石是由地质变化形成的，这些地质变化包括地壳变迁运动、火山爆发、沉积作用、冰川作用和风化作用等。在上述地质变化的作用下，生成了火成岩等。上述这些岩石中，有些处于共生状态——这也是自然界没有纯净矿石的主要原因。人们之所以需要分离技术与器械，即选矿装置与设备，其原因盖出于此。迄今为止，分离技术与器械尚远不能满足冶金工业的需要，科学家们正全力以赴，利用激光分离机把冶金工业技术提高到一个新水平。

  由上例可看出，科技英语在词汇、语法、文体等方面有着不同于日常英语的特征。以下我们将在比较科技英语文体与日常英语文体、文学文体的差异的基础上，了解科技英语的语言特点。

2 科技英语文体与日常英语文体的比较

  与日常英语相比，科技英语讲究严谨和庄重，在措辞、句式、段落、语篇结构等方面存在很大差异。在措辞方面，科技英语表现出专业术语多、名词化结构多、正式庄重的词多、实意动词多和借用词多等特点。在句式方面，科技英语多使用一般现在时、被动语态和长难句。在段落方面，科技英语比较规范：多数情况下有概括段落内容的主题句，而且出现在段首，读者只要阅读第一句就能得知段意。在语篇结构上，科技英语注重逻辑上的连贯、思维上的准确和严密、表达上的清晰与精炼，以客观的风格陈述事实和揭示真理。综上所述，科技英语在文体上的特点可归纳为：逻辑连贯、思维准确严密、陈述客观、表达清晰精炼。

  阅读下面文章"Natural and Synthetic Rubber"（见表1-1），试从词汇、语法、修辞等方面比较日常英语与科技英语两种文体的区别。

Unit One　Historical Development of Science and Technology

表 1-1　日常英语与科技英语文体比较

Natural and Synthetic Rubber	
日常英语	科技英语
People **get** natural rubber from rubber trees as a white, milky liquid, **which is called** latex. They **mix** it with acid, and dry it, **and then they send it** to countries all over the world. As the rubber industry **grew**, people **needed** more and more rubber. They **started** rubber plantations in countries with hot, **wet weather conditions**, but these still could not **give enough** raw rubber to **meet the needs** of growing industry.	Natural rubber **is obtained** from rubber trees as a white, milky liquid **known as** latex. This **is treated with** acid and dried **before being dispatched** to countries all over the world. As the rubber industry **developed**, more and more rubber **was required**. Rubber plantations **were established** in countries with a hot, **humid climate**, but these still could not **supply sufficient** raw rubber to **satisfy the requirements** of developing industry.
It was **not satisfactory** for industry to depend on supplies, **which come** from so far away from the industrial areas of Europe. It was always possible that wars or shipping trouble could stop supplies.	It was **unsatisfactory** for industry to depend on supplies **coming** from so far away from the industrial areas of Europe. It was always possible that supplies could be stopped by wars or shipping trouble.
For many years **people tried to make something to take its place**, but they **could not do it. In the end**, they **found a way** of making **artificial, man-made** rubber which is in many ways **better than** and in some ways **not as good as** natural rubber. They make artificial, man-made rubber in factories by a complicated chemical process. It is usually cheaper than natural rubber.	For many years, **attempts were made to produce a substitute**, but they **were unsuccessful**. Finally, a **method was discovered** of producing **synthetic** rubber which is in many ways **superior** and in some ways **inferior to** natural rubber. Synthetic rubber is produced in factories by a complicated chemical process. It is usually cheaper than natural rubber.
Today, the world needs so much **rubber** that we use both natural and artificial rubber in large **amounts**.	At present, the world requirements for rubber are so great that both natural and synthetic rubber is used in quantities.

（黄忠廉，李亚舒，2004：162）

比较两种文体后可知：

1）措辞上，与日常英语相比，科技英语文体显得更正规、庄重，见表 1-2。

表 1-2　日常英语与科技英语在措辞上的比较

日常英语	科技英语
send	dispatch
take its place	substitute
better	superior
not as good as	inferior to
amounts	quantities
wet weather conditions	humid climate
artificial, man-made rubber	synthetic rubber

2）句式上，与日常英语相比，科技英语文体更多使用被动句式，更常使用非限定动词，尤其是分词，见表 1-3。

表 1-3　日常英语与科技英语在句式上的比较

日常英语	科技英语
People get natural rubber from rubber trees…	Natural rubber is obtained from rubber trees…
They mix it with acid, and dry it, and then they send it to…	This is treated with acid and dried before being dispatched to…
…people needed more and more rubber.	…more and more rubber was required.
They started rubber plantations in…	Rubber plantations were established in …
…people tried to make something to…	…attempts were made to…
…they found a way of…	…a method was discovered of…
which is called	known as
which come from	coming from

3）文体方面，与日常英语相比，科技英语文体结构更紧凑，逻辑性更强。

Unit One Historical Development of Science and Technology

3 科技英语文体与文学文体的比较

科技文献揭示的是物质世界，是反映客观事实和规律的信息。为了客观准确地描述客观世界的实情、传递科技信息，科技英语文体在语言使用上更规范地道，在语言形式上更简明准确，在语义理解上更直截明了。而像小说、诗歌之类的文学作品表现的主要是人类的精神世界，是人们对物质世界的感受而抒发出来的感情、思想。文学文体用于建构鲜明的文学形象，是反映主观世界的媒介，常常通过音韵和节奏、语调和句式、思想和情感的表达方式，来建立独特的人物、环境和情节，在语言使用上比较洒脱自如，因而常常违背常规的语法定式，超脱语言的表述形式，以突显语言的风格化和人物的鲜活性。科技英语文体与文学英语文体在文字运用上各有独特之处：科技文献追求的是形式与逻辑的结合；而文学作品追求的是形式与意境的统一。在修辞方面，科技文献以交际修辞为主，文风质朴，描述准确；而文学作品富于美学修辞手段和艺术色彩。科技文献的语言表现出逻辑的连贯性和内容的统一性，突显出科技文体语言简洁、事理明了的特征。

从以上分析可知，科技英语和文学英语在语言、文字运用和修辞上存在较大的区别。阅读例3和例4，试比较科技英语文体与文学英语文体的区别。

George stood out in fair sight, on the top of the rock, as he made his declaration of independence; the glow of dawn gave a flush to his swarthy cheek, and bitter indignation and despair, gave fire to his dark eye; and, as if appealing from man to the justice of God, he raised his hand to heaven as he spoke.

分析：本段摘自《汤姆叔叔的小屋》。本段的文学文体明显，用了明喻一则：as if appealing from man to the justice of God。此外，文学作品有时需要用一些夸张的手法来表述人的主观思想和情感，三个词 indignation、despair、appealing 充分显示出文学描述的特点，给读者强大的冲击力和感召力。这种冲击力量越大，就越能震撼读者。

People sometimes ask "why waste time studying fossils".

Studying fossils is far from being a waste of time, because many useful facts can be learned from them. Generally speaking, fossils are especially important because they are the only clues to the existence and appearance of life on the earth millions of years ago. When arranged in proper geological order, fossils reveal how life steadily developed from lowly-organized primeval creatures to the complex animals of today. Such knowledge helps us better to comprehend the origins and evolution of life, and this in turn helps us to understand a little of what we ourselves are.

Since different fossil types are found in different strata, certain distinctive fossils can be used to identify different kinds of sedimentary rocks, including those below ground level of separated by miles of ocean. Even rocks at very great depths, when bored by the drills of engineers, can be identified by their fossil contents. Fossils are therefore valuable aids to mining and petroleum engineers.

分析： 本篇为科普文章，其科技文体特征明显：全文除了一个 we 作主语外，没有再用任何人称代词，也没有用任何明喻或暗喻；文章被动语态较多，显示其客观性强；语法规范，更易于读者接受信息。通过阅读科技文体，我们就能感受到文章朴实无华的特质。

4 科技英语文体的特点

科技英语文体包括科普文章、科技论文、科技报道、实验指示、实验报告以及科技发展的历史等。科技英语文体种类不同，因此具有各自的特点，主要差异表现在内容、措辞、句式、修辞、结构等方面。如科普作品以语言生动、形式活泼为特色；科技论文和各种学术文献说理充分，理论鲜明，论述缜密，极具严谨典雅的风格；实验指导书结构简单，句子短小精悍；产品说明书的风格通俗易懂，朴实简明；专利文献内容专深，措辞周密，说明详细，句式固定，结构规范。尽管如此，科技文体仍具有一些共同点，即条理分明，层次清楚，语言简练，合乎逻辑。

因此，我们在阅读科技文章时，首先要注意文章的逻辑关系，准确把握

Unit One Historical Development of Science and Technology

该文体的结构特征和文章的语言逻辑。其次要注意文章的语篇衔接连贯。语篇衔接通过词汇或语法手段使文脉相通,形成语篇的有形网络。语篇连贯以信息发出者和接受者双方共同了解的情景为基础,通过推理来达到语义的连贯,这是语篇的无形网络。充分利用语篇的叙事次序(时间顺序、空间顺序等)和逻辑连接词有助于我们对文章的整体把握,做到传意达旨。

试分析以下科技英语文体的特点。

Food quickly spoils and decomposes if **it** is not stored correctly. Heat and moisture encourage the multiplication of microorganisms, and sunlight can destroy the vitamins in such food as milk. Therefore, most food should be stored in a cool, dark, dry place which is also clean and well ventilated.

Food that decomposes quickly, such as meat, eggs, and milk should be stored in a temperature of 5ºC-10ºC. In **this** temperature range, the activity of microorganisms is considerably reduced. In warm climates, this temperature can be maintained only in a refrigerator or in the underground basement of a house. In Britain, for six months of the year at least, **this** temperature range will be maintained in an unheated room that faces the north or the east. Such a room will be ideal for food storage during the winter months provided that **it** is well ventilated.

分析: 这是一篇科普文章。科普文章是普及科学知识的读物,它包罗万象,涉及天文、地理、物理、化学、生物、医学等方面的知识;其形式也多样化,有科学小丛书、百科全书、科普文摘、科学史、科学家传记等。科普文章内容上着重常识性、知识性和趣味性;语言上通俗易懂,深入浅出,语句简短,多用普通词汇。

从文章的逻辑关系来看,这篇文章组织严密,逻辑性强。文章的层次清楚,第一段第一句话概括了该段的中心思想,第二段具体说明了如何才能保证保存食物的要求。文章中还使用了词汇手段"词语的复现关系",如 food、decompose、store(storage)都出现两次以上;以意义相同或相近的词组的形式出现,如文章前面用 spoil、decompose,后面扩展为 the multiplication of microorganisms、the activity of microorganisms、destroy the vitamin 等;

前面用 cool，后面用各种方式表达同一概念：in a temperature of 5°C-10°C, in a refrigerator, in the underground basement of a house, in an unheated room that faces the north or the east, during the winter months。通过词汇在意义上的衔接把全篇文章的各部分紧紧地联系在一起，使文章结构紧凑，前后呼应。从文章的语篇衔接连贯角度来看，在衔接方面，使用语法手段的"照应"，第一段代词 it 指代 food，第二段 it 指代 room，第二段中 this 指代"温度范围"。

通过以上讲座，我们对科技英语文体在内容、措辞、句式、修辞、结构等方面的特点有了比较全面的了解，但真正要理解科技文章还需进行大量的实践。

Analyze the features of the following passages.

1

First, a long glass tube was taken. The tube was closed at the top and was then completely filled with water. Next it was placed vertically in a large barrel half-full of water. When the bottom of the tube was opened, the water level in the tube fell to a height of approximately 10 meters above the water level in the barrel. As a result, a vacuum was left in the upper part of the tube. The water in the tube was supported by the atmospheric pressure. The height of the column of water could therefore be used to measure atmospheric pressure.

2

Many parts of this machine are made of flammable plastic. Never place hot or burning objects on or near the washing machine.

When disconnecting the power cord from the power outlet, always take hold of the plug, and not the wire, and pull free. Never connect or disconnect the power plug with wet hands since you may receive an electric shock.

For really dirty clothing use hot water 40℃.

For removal of blood stains use cold water only.

STAIN REMOVAL AND BLEACHING

- Add 1/2 cap per liter of water.
- Soak laundry well in solution for at least 20-30 minutes and wash.
- Rinse thoroughly.

Power source: 220V/50Hz

Power consumption: 400W

Washing capacity: 3kg

Spin capacity: 3kg

Water supply pressure: $0.3kg/cm^2$-$10kg/cm^2$

Net weight: 30kg

Dimension: $500 \times 500 \times 850$mm

Part B Reading

Technological Evolution of Humankind

This article is about the topic of technology in human history. Technology is the making, modification, usage, and knowledge of tools, machines, techniques, crafts, systems, methods of organization, in order to solve a problem, improve a preexisting solution to a problem, achieve a goal or perform a specific function. It can also refer to the collection of such tools, machinery, modifications, arrangements and procedures. Technology significantly affects humans' ability to control and adapt to their natural environments. Its evolution has experienced a long history…

Paleolithic [1] (2.5 million–10,000 B.C.)

The use of tools by early humans was partly a process of discovery, partly of evolution. Early humans evolved from a species of foraging hominids which were already bipedal, with a brain mass approximately one third that of modern humans. Tool use remained relatively unchanged for most of early human history, but approximately 50,000 years ago, a complex set of behaviors and tool use emerged, believed by many archaeologists to be connected to the emergence of fully modern langu-age.

Human ancestors have been using stone and other tools since long before the emergence of Homo sapiens approximately 200,000 years ago. The earliest methods of stone tool making, known as the Oldowan [2] "industry", date back to at least 2.3 million years ago, with the earliest direct evidence of tool usage found in Ethiopia within the Great Rift Valley, dating back to 2.5 million years ago. This era of stone tool use is called the Paleolithic, or "Old Stone Age", and spans all of human history up to the development of agriculture approximately 12,000 years ago.

Unit One Historical Development of Science and Technology

To make a stone tool, a "core" of hard stone with specific flaking properties was struck with a hammerstone. This flaking produced a sharp edge on the core stone as well as on the flakes, either of which could be used as tools, primarily in the form of choppers or scrapers. These tools greatly aided the early humans in their hunter-gatherer lifestyle to perform a variety of tasks including butchering carcasses (and breaking bones to get at the marrow); chopping wood; cracking open nuts; skinning an animal for its hide; and even forming other tools out of softer materials such as bone and wood.

The earliest stone tools were crude, being little more than a fractured rock. In the Acheulian era, beginning approximately 1.65 million years ago, methods of working these stone into specific shapes, such as hand axes emerged. The Middle Paleolithic, approximately 300,000 years ago, saw the introduction of the prepared-core technique, where multiple blades could be rapidly formed from a single core stone. The Upper Paleolithic, beginning approximately 40,000 years ago, saw the introduction of pressure flaking, where a wood, bone, or antler punch could be used to shape a stone very finely.

The discovery and utilization of fire, a simple energy source with many profound uses, was a turning point in the technological evolution of humankind. The exact date of its discovery is not known; evidence of burnt animal bones at the Cradle of Humankind suggests that the domestication of fire occurred before 1,000,000 B.C.; scholarly consensus indicates that Homo erectus had controlled fire by between 500,000 B.C. and 400,000 B.C. Fire, fueled with wood and charcoal, allowed early humans to cook their food to increase its digestibility, improving its nutrient value and broadening the number of foods that could be eaten.

Other technological advances made during the Paleolithic era were clothing and shelter; the adoption of both technologies cannot be dated exactly, but they were a key to humanity's progress. As the Paleolithic era progressed, dwellings became more sophisticated and more elaborate; as early as 380,000 B.C., humans were constructing temporary wood huts. Clothing, adapted from the fur and hides of hunted animals, helped humanity expand into colder regions; humans began to migrate out of Africa by 200,000 B.C. and into other continents, such as Eurasia.

Neolithic[3] through classical antiquity (10,000 B.C.–300 A.D.)

Man's technological ascent began in earnest in what is known as the Neolithic period ("New Stone Age"). The invention of polished stone axes was a major advance because it allowed forest clearance on a large scale to create farms. The discovery of agriculture allowed for the feeding of larger populations, and the transition to a sedentist lifestyle increased the number of children that could be simultaneously raised, as young children no longer needed to be carried, as was the case with the nomadic lifestyle. Additionally, children could contribute labor to the raising of crops more readily than they could to the hunter-gatherer lifestyle.

With this increase in population and availability of labor came an increase in labor specialization. What triggered the progression from early Neolithic villages to the first cities, such as Uruk, and the first civilizations, such as Sumer, is not specifically known; however, the emergence of increasingly hierarchical social structures, the specialization of labor, trade and war amongst adjacent cultures, and the need for collective action to overcome environmental challenges, such as the building of dikes and reservoirs, are all thought to have played a role.

Continuing improvements led to the furnace and bellows and provided the ability to smelt and forge native metals (naturally occurring in relatively pure form). Gold, copper, silver, and lead, were such early metals. The advantages of copper tools over stone, bone, and wooden tools were quickly apparent to early humans, and native copper was probably used from near the beginning of Neolithic times (about 8000 B.C.). Native copper does not naturally occur in large amounts, but copper ores are quite common and some of them produce metal easily when burned in wood or charcoal fires. Eventually, the working of metals led to the discovery

Unit One   Historical Development of Science and Technology

of alloys such as bronze and brass (about 4000 B.C.). The first use of iron alloys such as steel dates to around 1400 B.C.

Meanwhile, humans were learning to harness other forms of energy. The earliest known use of wind power is the sailboat. The earliest record of a ship under sail is shown on an Egyptian pot dating back to 3200 B.C. From prehistoric times, Egyptians probably used the power of the Nile annual floods to irrigate their lands, gradually learning to regulate much of it through purposely built irrigation channels and "catch" basins. Similarly, the early peoples of Mesopotamia [4], the Sumerians, learned to use the Tigris and Euphrates rivers for much the same purposes. But more extensive use of wind and water (and even human) power required another invention.

According to archaeologists, the wheel was invented around 4000 B.C. probably independently and nearly-simultaneously in Mesopotamia (in present-day Iraq), the Northern Caucasus (Maykop culture) and Central Europe. Estimates on when this may have occurred range from 5500 to 3000 B.C., with most experts putting it closer to 4000 B.C. The oldest artifacts with drawings that depict wheeled carts date from about 3000 B.C.; however, the wheel may have been in use for millennia before these drawings were made. There is also evidence from the same period of time that wheels were used for the production of pottery. More recently, the oldest-known wooden wheel in the world was found in the Ljubljana [5] marshes of Slovenia [6].

The invention of the wheel revolutionized activities as disparate as transportation, war, and the production of pottery (for which it may have been first used) . It didn't take long to discover that wheeled wagons could be used to carry heavy loads and fast (rotary) potters' wheels enabled early mass production of pottery. But it was the use of the wheel as a transformer of energy (through water wheels, windmills, and even treadmills) that revolutionized the application of nonhuman power sources.

Medieval and modern history (300 A.D.–)

An innovation continued through the Middle Ages with innovations such as silk, the horse collar and horseshoes in the first few hundred years after the fall of the Roman Empire. Medieval technology saw the use of simple machines (such as the lever, the screw, and the pulley) being combined to form more complicated tools, such as the wheelbarrow, windmills and clocks. The Renaissance brought forth many of these innovations, including the printing press (which facilitated the greater communication of knowledge), and technology became increasingly associated with science, beginning a cycle of mutual advancement. The advancements in technology in this era allowed a more steady supply of food, followed by the wider availability of consumer goods.

Starting in the United Kingdom in the 18th century, the Industrial Revolution was a period of great technological discovery, particularly in the areas of agriculture, manufacturing, mining, metallurgy and transport, driven by the discovery of steam power. Technology later took another step with the harnessing of electricity to create such innovations as the electric motor, light bulb and countless others. Scientific advancement and the discovery of new concepts later allowed for powered flight, and advancements in medicine, chemistry, physics and engineering. The rise in technology has led to the construction of skyscrapers and large cities whose inhabitants rely on automobiles or other powered transit for transportation. Communication was also greatly improved with the invention of the telegraph, telephone, radio and television. The late 19th and early 20th centuries saw a revolution in transportation with the invention of the steam-powered ship, train, airplane, and automobile.

The 20th century brought a host of innovations. In physics, the discovery of nuclear fission has led to both nuclear weapons and nuclear power. Computers were also invented and later miniaturized utilizing transistors and integrated circuits. The technology behind got called information technology, and these advancements subsequently led to the creation of the Internet, which ushered in the current Information Age. Humans have also been able to explore space with satellites (later used for telecommunication) and in manned missions going all the way to the moon. In medicine, this era brought innovations such as open-heart surgery and later stem cell therapy along with new medications and treatments. Complex manufacturing and construction techniques and organizations are needed to construct and maintain these new technologies, and entire industries have arisen to support and develop succeeding generations of increasingly more complex tools. Modern technology increasingly relies on training and education—their designers, builders, maintainers, and users often require sophisticated general and specific training. Moreover, these technologies have become so complex that entire fields have been created to support them, including engineering, medicine, and computer science, and other fields have been made more complex, such as construction, transportation and architecture.

New Words and Expressions

forage 采食，寻食
hominid 原人，原始人类
bipedal 两足动物的
archaeologist 考古学家
Homo sapiens 人类；现代人
Ethiopia 埃塞俄比亚（东非国家）
Great Rift Valley 东非大裂谷
span 横跨，跨越
flake 石片
chopper 斧头
scraper 刮刀；削刮器
butcher 屠宰

carcass（动物的）尸体
marrow 骨髓
fracture（使）破裂，（使）断裂
Acheulian 阿舍利文化的
antler 鹿角
domestication 驯养，驯化
consensus（意见等的）一致
migrate 迁移，移居
Eurasia 欧亚大陆
ascent 提高，升高
nomadic 游牧的，流浪的
hierarchical 等级制度的，等级体系的

dike 堤，坝；沟，渠，排水道
bellows 风箱
bronze 青铜
brass 黄铜
channel 沟渠
basin 流域；盆地
Sumerian 苏美尔人
Tigris 底格里斯河
Euphrates 幼发拉底河
Caucasus 高加索山脉
marsh 沼泽，湿地
disparate 不能比拟的
wagon 运货马车
treadmill（古时罚囚犯踩踏的）踏车
lever 杠杆
screw 螺钉
pulley 滑轮
wheelbarrow 独轮小车，手推车
metallurgy 冶金术
transit 运输线；公共交通系统
usher 展示；开创

Notes

1. Paleolithic 旧石器时代。人类以石器为主要劳动工具的早期泛称旧石器时代，从距今 260 万年延续到 1 万多年以前。

2. Oldowan（东非旧石器时代）奥尔德沃文化的。该文化以简单的石制砍砸器、刮削器为特征。

3. Neolithic 新石器时代。始于距今 8000 年前的人类原始（母系）氏族的繁荣时期，以磨制的石斧、石锛、石凿和石铲、琢制的磨盘和打制的石锤、石片、石器为主要工具。

4. Mesopotamia 美索不达米亚。指底格里斯与幼发拉底两河的中下游地区。

5. Ljubljana 卢布尔雅那（斯洛文尼亚共和国首都）。

6. Slovenia 斯洛文尼亚（前南斯拉夫联邦成员共和国，1991 年宣布独立，位于欧洲）。

Unit One Historical Development of Science and Technology

Exercises

I. Find out the English equivalents to the following Chinese terms from the passage.

1. 旧石器时代
2. 考古学家
3. 依靠狩猎和采集的生活方式
4. 人类的发源地
5. 新石器时代
6. 游牧生活方式
7. 等级制度的社会结构
8. 史前古器物
9. 四轮的运货马车
10. 马轭
11. 印刷机
12. 核分裂
13. 集成电路
14. 信息时代
15. 干细胞治疗

II. Choose the best answer according to the information in the passage.

1. According to most archaeologists, what is connected with the appearance of modern language?

    A) The emergence of modern tools.

    B) The emergence of a complex set of behaviors and tool use.

    C) The emergence of stone and ancient tools.

    D) The emergence of early humans.

2. What is the feature of tool making in the Acheulian era?

    A) Making stones into specific shapes.

    B) The appearance of the prepared-core technique.

    C) The introduction of pressure flaking.

    D) Shaping a stone very finely.

3. What is the significance of utilization of fire to humans?

    A) Making food more digestive.

    B) Improving the food's nutrient.

    C) Giving humans more choices on food.

D) All of the above.

4. Why was the invention of polished stone axes a major advance in Neolithic period?

   A) It allowed for the feeding of larger populations.

   B) It improved the fineness of stone tools.

   C) It made possible that children contribute labor to hunter-gatherer life.

   D) It allowed forest clearance on a large scale to create farms.

5. Which one is NOT the achievement in the Industrial Revolution?

   A) The discovery of steam power.

   B) The invention of the wheel.

   C) The electric motor.

   D) The automobiles.

Part C Extended Reading

Invention of Wheel

Evidence of wheeled vehicles appears from the mid-4th millennium B.C., near-simultaneously in Mesopotamia, the Northern Caucasus and Central Europe, so that the question of which culture originally invented the wheeled vehicle remains unresolved and under debate.

The earliest well-dated depiction of a wheeled vehicle (here a wagon—our wheels, two axles), is on the Bronocice pot, a ca. 3500–3350 B.C. clay pot excavated in a Funnelbeaker culture settlement in southern Poland.

The wheeled vehicle spread from the area of its first occurrence (Mesopotamia, Caucasus, Balkans, and Central Europe) across Eurasia, reaching the Indus Valley by the 3rd millennium B.C. During the 2nd millennium B.C., the spoke-wheeled chariot spread at an increased pace, reaching both China and Scandinavia by 1200 B.C. In China, the wheel was certainly present with the adoption of the chariot in ca. 1200 B.C., although Barbieri-Low argues for earlier Chinese wheeled vehicles, circa 2000 B.C.

Although they did not develop the wheel proper, the Olmec and certain other Western Hemisphere cultures seem to have approached it, as wheel-like worked stones have been found on objects identified as children's toys dating to about 1500 B.C. It is thought that the primary obstacle to large-scale development of the wheel in the Western Hemisphere was the absence of domesticated large animals which could be used to pull wheeled carriages. The closest relative of cattle present in Americas in pre-Columbian times, the American Bison, is difficult to domesticate and was never domesticated by Native Americans; several horse species existed until about 12,000 years ago, but ultimately went extinct. The only large animal that was domesticated in the Western Hemisphere, the llama, did not spread far beyond the Andes by the time of the arrival of Columbus.

Early antiquity Nubians used wheels for spinning pottery and as water wheels. It is thought that Nubian waterwheels may have been ox-driven. It is also known that Nubians used horse-driven chariots imported from Egypt.

The invention of the wheel thus falls in the late Neolithic, and may be seen in conjunction with other technological advances that gave rise to the early Bronze Age.

Wide usage of the wheel was probably delayed because smooth roads were needed for wheels to be effective. Carrying goods on the back would have been the preferred method of transportation over surfaces that contained many obstacles. The lack of developed roads prevented wide adoption of the wheel for transportation until well into the 20th century in less developed areas.

Early wheels were simple wooden disks with a hole for the axle. Because of the structure of wood, a horizontal slice of a tree trunk is not suitable, as it does not have the structural strength to support weight without collapsing; rounded pieces of longitudinal boards are required. The oldest known example of a wooden wheel and its axle were found in 2003 in the capital of Slovenia. According to the radiocarbon dating, it is between 5100 and 5350 years old. It has a diameter of 72 centimeters (28 inches) and has been made of ash wood, whereas its axle has been made of oak.

The spoked wheel was invented more recently, and allowed the construction of lighter and swifter vehicles. In the Harappan civilization of the Indus Valley and Northwestern India, we find toy-cart wheels made of clay with spokes painted or in relief, and the symbol of the spoked wheel in the script of the seals, already in the second half of the 3rd millennium B.C. The earliest known examples of wooden spoked wheels are in the context of the Andronovo culture, dating to ca 2000 B.C. Soon after this, horse cultures of the Caucasus region used horse-drawn spoked-wheel war chariots for the greater part of three centuries. They moved deep into the Greek peninsula where they joined with the existing Mediterranean peoples to give rise, eventually, to classical Greece after the breaking of Minoan dominance and consolidations led by pre-classical Sparta and Athens. Celtic chariots introduced an iron rim around the wheel in the 1st millennium B.C. The spoked wheel was in continued use without major modification until the 1870s, when wire wheels and pneumatic tires were invented.

The invention of the wheel has also been important for technology in general, important applications including the water wheel, the cogwheel, the spinning wheel, and the astrolabe or torquetum. More modern descendants of the wheel include the propeller, the jet engine, the flywheel (gyroscope) and the turbine.

New Words and Expressions

axle 车轴
spoke 刹车；辐条
Scandinavia 斯堪的那维亚半岛
Olmec 奥尔梅克文化的
Bison 北美野牛；欧洲野牛
llama 美洲驼，无峰驼
Andes 安第斯山脉
Nubians 努比亚人
disk 圆盘，轮盘
slice 板；块
longitudinal 纵向的
ash 桉树

consolidation 巩固，加强
Sparta 斯巴达（希腊南部的古代城邦）
Celtic 凯尔特族的
rim 边；缘；轮辋
pneumatic 空气的；气动的
cogwheel 钝齿轮，木齿铁轮
astrolabe 古代的天体观测仪，星盘
torquetum 赤基黄道仪
propeller 螺旋桨；推进器
gyroscope 回转仪；陀螺仪
turbine 涡轮（机），叶轮（机）

Exercises

I. Find out the English equivalents to the following Chinese terms from the passage.

1. 陶罐
2. 北美野牛
3. 马拉战车
4. 青铜时代
5. 树干
6. 结构力
7. 放射性碳年代测定法
8. 希腊半岛
9. 气胎
10. 水车
11. 手纺车
12. 喷气式发动机
13. 飞轮

II. Answer the following questions by using the sentences in the passage or in your own words.

1. According to the passage, which culture originally invented the wheeled vehicle?
2. What was the primary obstacle to large-scale development of the wheel in the Western Hemisphere?
3. What prevented wide adoption of the wheel for transportation before the 20th century?
4. What's the result of the inventions of wire wheels and pneumatic tires?
5. What does the word "descendants" in the last paragraph specifically mean?

Unit Two
Ancient Civilizations

Part A Lecture

提高科技英语阅读理解能力

在科学技术飞速发展、世界面貌日新月异的今天，阅读是人们获得科学发展最新信息的主要途径。为了及时掌握最新的科技信息，把握现代科技发展的脉搏，我们应该重视阅读能力的培养。较强的阅读能力表现在能迅速无误地从一段文字中获取最大限量的信息。要从浩如烟海的外语资料中获得有用的科技信息，我们需要有较强的科技英语阅读理解能力，而这需要通过学习和阅读实践才能获得。本单元将阐述科技英语阅读理解的内涵，分析影响科技英语阅读理解的因素，介绍提高科技英语阅读理解能力的方法。

1 科技英语阅读与理解能力的内涵

"阅读"和"理解"是两个密切相关、但又不尽相同的概念。"阅读"是对文字和句子表层意义进行识别的过程，主要需要相应的词汇、语法知识；"理解"是对文章所传达信息的获取，是把读者头脑中的知识与文章中的知识相互作用，再把文章中的知识"内化"成读者智能的过程，是一个由感性认识上升到理性认识的过程，主要需要英语语篇知识、超越句子水平的阅读能力和逻辑分析能力（比如推理、判断、概括、归纳、总结等），还需要社会文化背景知识等。阅读理解是从书面语言中获取信息的一种复杂的智力活动，是从书面语言中获得意义的心理过程，在这些过程中表现出来的能力，被称为"阅读能力"或"阅读理解能力"。科技英语文献阅读的目的是为了理解科技文献的内容并获取所需的信息。科技英语阅读理解能力主要与语言知识、篇章能力、知识结构和阅读技能等因素相关。

Unit Two　Ancient Civilizations

1·1 语言知识

扎实的语言知识是科技英语阅读的基础。首先要掌握一定的词汇量，尤其是科技术语或专业词汇。词汇量是顺利阅读的前提。词汇量不足往往会造成理解上的失误。如：

Buffer **memory** read/write operations and **gate** opening/closing are executed at times defined by the connection to be established through the **switching network**.

误译：缓冲记忆读/写操作及门的打开/闭合的执行时间由开关网络需建立起的接续来规定。

分析：由于原文中的专业词汇被简单地理解成普通词汇，所以这里有三处理解失误：一是 memory，此处应是"存储器"，而不应该理解为普通英语中的"记忆"；二是 gate，此处是指"门电路"，而不应该理解为常见意义"门"；三是 switching network，应是"交换网络"的意思，而不是"开关网络"。

其次是掌握英语的语法和习惯用法。牢固掌握语法后，才能在阅读中迅速掌握各种语言形式所传递的确切信息，在遇到结构复杂的句子时，还可借助语法分析的方法来理解其意。

The advent of stored program control in telephone exchanges **has had as great an effect on** exchange administration and maintenance **as** on call processing.

误译：储存程序控制的出现在电话交换机中已对交换机的管理与维护产生巨大影响，像呼叫处理一样。

分析：由于没有正确理解 have a great effect on 和 as...as... 结构的结合，造成理解上的偏差。例句的意思为：在电话交换机中，储存程序控制技术的出现对交换机管理与维护的影响之大，如同它对呼叫处理一样。

从以上两例可以看出，理解上的偏差可能出现在词语层面（例1），也可能出现在对句法的理解上（例2）。因此，打好语言知识的基础有助于提高阅读科技英语文献的速度和理解的准确度。

27

1.2 篇章能力

篇章能力是有效阅读必不可少的一个因素，是科技英语阅读理解能力的重要组成部分。篇章能力指理解比句子更大的语言单位的能力，包括：1）体现语篇表层结构的语法衔接手段（如照应、替代、省略等）和词汇衔接手段（如复现关系、同现关系等）；2）使语篇语义关联的连贯；3）句组、段落和语篇的构成和特点；4）句际间和段际间的关系（如并列关系、转折关系、因果关系等）。较强的篇章能力可以帮助我们厘清语篇的结构，抓住主题，弄清作者的思路、意图、目的，从而把握全局，提高理解的深度和广度。如果没有理清语篇的结构，就会出现理解上的偏差。如：

Energy is always leaking away and taking forms in which we cannot catch it and use it. The fire under your boiler heats other things besides the water in the boiler. The steam seeps out around the piston. The air around the hot cylinder of the steam engine warms up and blows away.

误译：能量总是在不断地流失，而且它所出现的形式，往往也使我们难以捕捉和利用。你的锅炉底下的火除了加热锅炉里的水以外，还加热了一些其他的东西。蒸汽从活塞周围的缝隙渗漏出来。围绕蒸汽机灼热汽缸的空气被加热、带走。

分析：实际上，原文中的四句话是密切联系在一起的句群，第1句作结论，第2、3、4句话举例加以证明。上文将四句话独立开来理解，没有将四句话之间暗含的内在联系体现出来。本例句可理解为：能量总是以我们无法驾驭和利用的形式在不断地流失。比如，锅子底下的火除了加热锅中的水以外，也还加热了一些其他的东西；蒸汽从活塞周围的缝隙渗漏掉；蒸汽机灼热汽缸外围的空气被加热后散发出去。

因此，要正确理解原文，有必要依据上下文关系吃透原文词义，辨明原文句法语法关系，厘清原文的篇章结构和语句段落之间的联系。

1.3 知识结构

知识结构对科技英语的阅读理解有至关重要的影响。阅读是思想与语言相互作用的过程，需要利用各种知识去理解读物所传达的信息。这些知识包括背景知识和专业知识。

Unit Two Ancient Civilizations

　　背景知识包括不同国家、不同地区、不同民族、不同阶层人士的思维方式、语言表达方式、风俗、习惯和心态，也包括不同社会制度和不同文化的广博知识。

　　The technologies that are particularly dangerous over the next hundred years are nanotechnology, artificial intelligence and biotechnology. The benefits they will bring are beyond doubt but they are potentially very dangerous. In the field of artificial intelligence there are prototype designs for something that might be 50,000 million times smarter than the human brain by the year 2010. The only thing not feasible in the film *Terminator* is that the people win. If you're fighting against atechnology that is much smarter than you, you probably will not win. We have all heard of the **grey goo** problem that self-replicating nanotech devices might keep on replicating until the world has been reduced to sticky goo, and certainly in biotechnology, we have really got a big problem because it is converging with nanotechnology. Once you start mixing nanotech with organisms and you start feeding nanotech-enabled bacteria, we can go much further than **the Borg in *Star Trek***, and those superhuman organisms might not like us very much.

　　分析：本段中的"grey goo（灰色黏质）"，"the Borg（博尔族人）in *Star Trek*"与小说 *Engines of Creation* 以及美国一部曾风靡一时的科幻电影《星际旅行》有关。小说和电影情节等背景知识有助于我们更深入地理解文章，形成正确的观点。

　　专业知识指厚实的专业基础。常言道，隔行如隔山。科技英语文献大多专业性较强，具有专业术语多、信息量大的特点。为了顺利进行阅读，学习者的专业水平也有一定影响。专业知识丰富，有利于进行阅读；反之，专业知识不足，就会在阅读中产生某些困难。阅读科技文献要求阅读者能够对包括各种专门术语、概念在内的信息进行系统的理解、分类、记忆，掌握大量技术词汇，学会运用公式和定律等。

　　When a new type of aero-plane is being made, normally only one of the first three made will be flown. Two will be destroyed on the ground in structural tests. The third one will be tested in the air.

分析：本段涉及新型飞机的三架试验机的用途。如果不了解飞机制造过程，就很难理解段中的"Two"是"两个"还是"第二"。航空航天方面专家[①]解释道：型号研制时，一般有 5 架实验飞机，至少是 3 架。01 号机用于静电实验，在地面进行，用于受力测试，一般实验后这架飞机不可复用。02 号飞机用于试飞实验，在空中飞行，用于飞行各参数测试，数据较多，是对型号进一步改进的重要手段，可复用。03 号飞机用于飞机强度使用寿命等测试，也是改进型号的措施之一，不可复用。有时飞机实验中出说问题或不能使用或改进，可用 04 或 05 备用飞机代替。

由此可见三架实验机中有两架飞机将在进行的结构实验中遭到破坏。根据专业知识，本段可理解为：当制造一种新型飞机的时候，在首先制造出来的三架飞机中，通常只有一架飞机能飞行。两架飞机在地面上进行的结构实验中被破坏。第三架飞机则在空中实验。

从例 5 可知，专业知识的运用在科技英语文献阅读中的地位不容忽视，如果专业知识太过缺乏，就会导致理解的不准确，甚至让人产生误会。

总之，拥有背景知识和专业知识有助于我们更透彻地理解文章内容和作者意图，从而提高科技英语阅读能力。

1·4 阅读技能

阅读技能包括确定文章的主题思想以及对围绕和支持该主题思想的若干条次级信息（比如细节、事实、观点等）的理解，猜测词义，进行判断和推理，把握字里行间的言下之意，确定作者观点和态度；也包括阅读策略的能力（即利用一切可以利用的线索补足读物所缺信息的能力）。

例 6

What has the telephone done to us, or for us, in the hundred years of its existence? A few effects suggest themselves at once. It has saved lives by getting rapid word of illness, injury, or fire from remote places. By joining with the elevator to make possible the multi-story residence or office building, it has made possible—for better or worse—the modern city. By

---

① 南京航空航天大学宇航学院童明波教授。

Unit Two　Ancient Civilizations

bringing about a great leap in the speed and ease with which information moves from place to place, it has greatly accelerated the rate of scientific and technological changes and growth in industry. Beyond doubt it has seriously weakened if not killed the ancient art of letter writing. It has made living alone possible for persons with normal social impulses; by so doing, it has played a role in one of the greatest social changes of this century, the breakup of the multi-generational household. It has made the war chillingly more efficient than formerly. Perhaps, though not provably, it has prevented wars that might have arisen out of international misunderstanding caused by written communication. Or perhaps—again not provably—by magnifying and extending irrational personal conflicts based on voice contact, it has caused wars. Certainly it has extended the scope of human conflicts, since it impartially disseminates the useful knowledge of scientists and the nonsense of the ignorant, the affection of the affectionate and the malice of the malicious.

(1) What is the main idea of this passage?
　　A) The telephone has helped to save people from illness and fire.
　　B) The telephone has helped to prevent wars and conflicts.
　　C) The telephone has made the modern city neither better nor worse.
　　D) The telephone has had positive as well as negative effects on us.
(2) According to the passage, it is the telephone that _____.
　　A) has made letter writing an art
　　B) has prevented wars by avoiding written communication
　　C) has made the world different from what it was
　　D) has caused wars by magnifying and extending human conflicts
(3) The telephone has intensified conflicts among people because _____.
　　A) it increases the danger of war
　　B) it provides services to both the good and the malicious
　　C) it makes distant communication easier
　　D) it breaks up the multi-generational household
(4) The author describes the telephone as impartial because it _____.
　　A) saves lives of people in remote places

B) enables people to live alone if they want to

C) spreads both love and ill will

D) replaces much written communication

(5) The writer's attitude towards the use of the telephone is _____.

A) affectionate   B) disapproving   C) approving   D) neutral

分析：

题 1 要求读者通过阅读全文确定文章的中心思想。通读全文后可知，电话的出现既给人类的生活带来了便利，也为战争带来了便利，故答案为 D。

题 2 要求读者针对全文意思做出合理的推论，故答案为 C。

题 3 和题 4 要求读者从文章的相关部分找出事实依据，考核辨认重要事实的能力。题 3 的依据是文章倒数第二句话，故答案为 C。

题 4 的依据是文章的最后一句话，故答案为 C。

题 5 要求读者判断作者的态度，考核判断能力。文章叙述了电话在日常生活中的作用，既有积极的作用，也有消极的作用，故答案为 D。

从例 6 可知，读者想通过阅读获取信息的话，就需掌握阅读技能，包括从整篇文章层面上，也包括从句子层面上获取信息的能力，具体而言包括确定中心思想、辨认重要事实、判断与推理等能力。

以上 6 例告诉我们，科技英语阅读理解能力的获得有赖于多种因素，比如扩大积累词汇、掌握基本语法和习惯用法、理解语篇结构、了解英语民族社会文化背景知识、熟悉相关专业知识，掌握阅读技能等。

2 提高科技英语阅读理解能力的方法

通过分析和了解科技英语阅读理解能力的内涵，我们可以从以下几方面提高科技英语阅读理解能力。

2·1 掌握一定的词汇量

词汇是构筑句子和文章的基本要素，也是阅读文章时首先要理解的成分。我们可通过扩大词汇量、猜测词义、熟悉常用词汇的专业化用法等方法掌握词汇。

2·2 掌握语法规则

掌握语法规则，可以有效提高分析复杂句型的能力。我们知道，英语的语法手段和表达习惯与汉语不同，英语中有大量错综复杂的长句，大量语法功能很强的介词、连词，丰富的词性变化，各种短语和不同从句，以及许多语法手段。此外，主句中经常有从句，短语中有短语，短语和从句互相包含容纳，语序灵活。这使得英语句子层次繁杂，像一棵枝繁叶茂的大树，盘根错节，叠床架屋。尤其在科技英语文献中，为了说理严谨，逻辑紧密，描述准确，复杂的长难句大量出现。如：

The technical **possibility** could well exist, therefore, **of** nationwide integrated transmission network of high capacity, controlled by computers, inter-connected globally by satellite and submarine cable, providing speedy and reliable communications throughout the world.

分析：本句主句为 The technical possibility could well exist，其余是很长的 of 介词短语做 possibility 的定语，句子语义重心为：这种可能性从技术上来讲是完全存在的。本句译为：建立全国统一的大容量的通讯网络，由计算机进行控制，通过卫星和海底电缆在全球从事快速可靠的通讯服务，这种可能性从技术上来讲是完全存在的。

从以上例子看出，阅读科技英语长句时，首先要通读全句，厘清句法结构。语法关系不清必然导致理解上的错误。要正确划分主要成分和次要成分的范围和层次关系，注意长句子的连接手段，如连词、介词、非谓语形式（不定式短语、分词短语、动名词短语）、各种从句（主语从句、宾语从句、同位语从句、定语从句、状语从句）。理解语法关系之后要领会句子的要旨和逻辑关系，分清重心，突出重点。

2·3 识别逻辑衔接词

识别和依靠句子之间的逻辑衔接词，有助于弄清文章层次，超越句子水平理解文章意义并获取信息。逻辑衔接词可以表示多种不同的语义关系，包括递进关系、转折关系、因果关系、顺序关系和总结关系五大类。如：

It will no doubt be a revolution, **but** there are some scary *Brave New World* overtones that raise fundamental questions about how we will think about ourselves.

分析：例句中 but 的前一部分告诉我们"这一发现无疑是革命性的"，but 的后一部分说明"其中却蕴含着《勇敢的新世界》②式的令人恐慌的色彩，它给人们提出了一些有关我们将怎样认识我们自己的基本问题"，前后内容不一致，但通过衔接词 but 我们知道，后面的内容才是重点所在。

2.4 了解段落常用的展开方式

段落是由句子组成的语义单位，它既是文章的组成部分，又是自成一体、相对独立的整体。英语句子具有重形合、注重形式结构的特点，句子与句子之间的连接有一定的规律可循。英语的段落结构往往使段落中的其他句子一方面以语义与主题句直接关联，另一方面以一些逻辑衔接词显现逻辑序列和句际关系。因此，英语段落条理清晰，层次结构清楚，衔接自然流畅，句与句之间有内在的逻辑关系。

英文段落常用的展开方式有：因果分析法、比较对照法、例证法、分类法、定义法等。每一种展开方式都有其特定的逻辑衔接词。熟悉这些衔接词，了解段落的展开方式，对阅读理解大有裨益。

There is a difference between science and technology. **Science** is a method of answering theoretical questions; **technology** is a method of solving practical problems. **Science** has to do with discovering the facts and relationships between observable phenomena in nature and with establishing theories that serve to organize these facts and relationships; **technology** has to do with tools, techniques, and procedures for implementing the findings of science. Another distinction between **science** and **technology** has to do with the progress in each.

---

② 《勇敢的新世界》是英国作家 Aldous Huxley 的一部科幻小说，描述了一个人被基因控制的恐怖世界。

分析：本段使用比较对照法展开段落。第 1 句为主题句，说明"科学与技术之间有区别"，然后通过反复以科学和技术为各句的基本主题展开说明了区别之所在。

2·5 了解语篇衔接手段的使用

我们知道，词汇和词组构成句子，句子依据相关程度组成句群，句群构成段落。段落是语篇的基本单位，人们常把段落说成是表达一个完整思想的一组句子。所谓思想"完整"是相对而言的，有些段落表达的思想比较完整，而更多的段落虽然形式上自成一体，但也只是一个完整思想中的一部分。因此，段落与段落之间的关系既有独立性，又有依赖性。辨别段落与段落之间的关系有助于我们正确理解语篇的衔接。请看下例。

[1] The argument was finally settled a hundred years later in a famous series of experiments by Pasteur. Pasteur proved, once and for all, that bacteria are not generated spontaneously—any more than Van Helmont mice. He showed that bacteria are not the product of decay but the cause of decay. He communicated his results to the general public in a famous lecture which he gave at the University of Paris 92 years ago and in which he demonstrated three important experiments.

[2] In the first experiment, Pasteur showed that if you destroy the bacteria in a suitable medium by boiling, and allow only sterile air to enter the flask, you get no subsequent growth of bacteria.

[3] In his next experiment he proved that growth of bacteria would start in the medium if it was inoculated with dust collected from the air.

[4] The third experiment—and the one of which he was obviously most proud—was one in which he prepared his broth in an ordinary flask which he then pulled out in a flame so that it has gooseneck. He boiled the medium in the flask for three to four minutes, allowing the steam to go up the gooseneck. Then he simply turned off the burner and let the flask sit there until it cooled. Then, without sealing it, and without any other precaution, he put the flask in an incubator and left it there. Nothing grew in it. The flask was completely open to the air, and there was no question

of the oxygen being depleted because oxygen had free access to the flask. Yet the broth remained sterile. If you visit the Institute Pasteur in Paris today you will still see such flasks which, it is said, were put there by Pasteur.

[5] The explanation which Pasteur gave for this experiment is this: When the broth is boiled, the steam comes out of the gooseneck and, of course, drives out the air. When the flame is turned off, air reenters the flask, but it comes in contact with liquid which is almost at boiling point—hot enough to kill bacteria. As the broth cools down, the stream of air entering the flask slows down very much, to the point where dust particles in the air can no longer trip; they get caught in the moist gooseneck, so that they never reach the surface of the broth after it is cool.

分析：文章第一段讲述法国著名生物学家巴斯德证明细菌不是自己产生的，不是腐烂才产生细菌，而是细菌造成腐烂。该段有一定的独立性，但其对上下文的依赖性也很明显。第一句的"这一争论"当然是上文已经提到的，而且"一百年以后"也必须参考前文来计算出来。文章接下来分三段讲述三次实验。第二段讲述第一次实验；第三段是第二次实验；第四段是第三次实验。第五段与第四段密切相关，甚至可以合并为一段。第四段描写第三次实验的具体过程，从实验本身讲算是完整的，但还没有解释为什么瓶内东西没有腐烂。因此，第五段讲的是该实验的理论解释，是对第四段的说明。

通过对以上五段的分析，我们可以看到存在于段落之间的关系：并列关系和主从关系。有些段落支配着其他段落，有些段落则是对其他段落的说明、解释或发展。以上五段的关系表达见图 2-1。

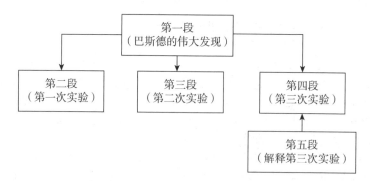

图 2-1 段落间的关系图

2·6 了解文章的逻辑结构

文章的基本结构是指文章的文脉，即句与句、段与段之间的逻辑关系，通过某条主线按逻辑构成一个整体。文章的主要逻辑结构有：1）时空顺序。按时间先后叙述某一事物，或说明某一事物由过去至现在的发展情况，或按空间顺序说明某一事物。2）总—分—总（总—分）顺序。首段进行总的概括，其他段落分别加以说明或具体论述首段的观点，末段再次总结。3）分—总顺序。前面几段分别说明观点，末段总结。4）对比。这类文章通常以对比各事物之间的共同点或差异为主题。

科技文章包括科普文章、科技论文、科技报道、科技发展史等。尽管种类不同，但科技文章仍具有一些共同点，即条理分明、层次清楚、语言简练、合乎逻辑。科技文章最常见的逻辑结构主要有"总—分—总"顺序。

"总—分—总"的逻辑结构分为三部分：1）引出要讨论的问题或现象（常以具体事例引出话题或论点）；2）然后展开分析过程或讨论其原因；3）最后得出结论、提出建议或解决方案。

2·7 掌握基本阅读技巧

阅读理解不是一种消极被动地接受文章信息的过程，而是一个积极主动地与文章作者交流并且获取信息的过程。在阅读理解时，阅读者要积极思考，着重掌握下列阅读技能：理解和确定文章的主旨大意；辨别主要事实与细节；根据上下文判断生词的意思；推敲文章和句子的言下之意；领会作者态度和文章语气；等等。我们将在本书的其他章节对这几种技能进行专题训练。

3 综合练习

本节将运用上文介绍的提高科技英语阅读理解能力的方法，分析下面短文的结构，并完成阅读理解题。

Shaping Factors of Inventions

[1] What accounts for the great outburst of major inventions in early America —breakthroughs such as the telegraph, the steamboat and the weaving machine?

[2] Among the many shaping factors, I would single out the country's excellent elementary schools; a labor force that welcomed the new technology; the practice of giving premiums（保险费）to inventors; and above all the American genius for nonverbal, "spatial" thinking about things technological.

[3] Why mention the elementary schools? Because thanks to these schools our early mechanics, especially in the New England and Middle Atlantic states, were generally literate and at home in arithmetic and in some aspects of geometry and trigonometry.

[4] Acute foreign observers related American adaptiveness and inventiveness to this educational advantage. As a member of a British commission visiting here in 1853 reported, "With a mind prepared by thorough school discipline, the American boy develops rapidly into the skilled workman."

[5] A further stimulus to invention came from the "premium" system, which preceded our patent system and for years ran parallel with it. This approach, originated abroad, offered inventors medals, cash prizes and other incentives.

[6] In the United States, multitudes of premiums for new devices were awarded at country fairs and at the industrial fairs in major cities. American flocked to these fairs to admire the new machines and thus to renew their faith in the beneficence of technological advance.

[7] Given this optimistic approach to technological innovation, the American worker took readily to that special kind of nonverbal thinking required in mechanical technology. As Eugene Ferguson has pointed out, "A technologist thinks about objects that cannot be reduced to unambiguous verbal descriptions; they are dealt with in his mind by a visual, nonverbal process…The designer and the inventor…are able to assemble and manipulate in their minds devices that as yet do not exist."

[8] This nonverbal "spatial" thinking can be just as creative as painting and

Unit Two Ancient Civilizations

writing. Robert Fulton once wrote, "The mechanic should sit down among levers, screws, wedges, wheels, etc., like a poet among the letters of the alphabet, considering them as an exhibition of his thoughts, in which a new arrangement transmits a new idea."

[9] When all these shaping forces—schools, open attitudes, the premium system, a genius for spatial thinking—interacted with one another on the rich U. S. mainland, they produced that American characteristic, emulation. Today that word implies mere imitation. But in earlier times it meant a friendly but competitive striving for fame and excellence.

(1) The term "spatial" (Paragraph 2) is related to _____.
   A) descriptions   B) process
   C) space   D) thoughts
(2) According to the author, the great outburst of major inventions in early America was in a large part due to _____.
   A) elementary schools
   B) enthusiastic workers
   C) the attractive premium system
   D) a special way of thinking
(3) It is implied that adaptiveness and inventiveness of the early American mechanics _____.
   A) benefited a lot from their mathematical knowledge
   B) shed light on disciplined school management
   C) was brought about by privileged home training
   D) owned a lot to the technological development
(4) A technologist can be compared to an artist because _____.
   A) they are both winners of awards
   B) they are both experts in spatial thinking
   C) they both abandon verbal description
   D) they both use various instruments
(5) The best title for this passage might be _____.
   A) Inventive Mind

B) Effective Schooling

C) Ways of Thinking

D) Outpouring of Inventions

分析：本文第二段共谈到了导致早期美国方面创造热潮的四个原因，接下来的几段进行了分别论述（并非一段谈一方面），最后一段又从这四方面进行了总结（注意：第四方面是作为最主要的因素提及的）。第二段为主旨句，提出四个因素（excellent elementary schools，a labor force，the American genius for nonverbal,"spatial" thinking about things technological）；第三段讲述第一方面 elementary schools；第四段讲述第二方面 skilled workman；第五、六段讲述第三方面 the "premium" system；第七、八段讲述第四方面 special kind of nonverbal thinking；最后一段结论，涵盖以上四个方面。

就以上问题来看，有的针对某个方面，有的针对整体提问。

第 1 题选项 C 是正确答案。根据构词法可以推测答案。

第 2 题选项 D 是正确答案。正如上文分析中指出的，在文章所谈到的四个方面的起因中，作者认为特殊的思维方式是主要的。注意提问部分用了 in a large part（主要的）。

第 3 题选项 A 是正确答案。本题显然提问的是第一方面；答案依据第四段第一句（过渡句）和第三段。

第 4 题选项 B 是正确答案，本题显然提问的是第四方面；参阅倒数第二段。

第 5 题选项 A 是正确答案。本题显然是一个综合题。本文用了很大篇幅描述四个起因，它主要是想说明的是：早期美国人所具有的创造力的动因。参阅第一段和最后一段。

Quiz

Read the following passage, and choose the best answer to each question.

The word *science* is heard so often in modern times that almost everybody has some notion of its meaning. On the other hand, its definition is difficult for many people. The meaning of the term is confused, but everyone should understand its meaning and objectives. Just to make the explanation as simple as possible, suppose science is defined as classified knowledge (facts).

Even in the true sciences, distinguishing fact from fiction is not always easy. For this reason, great care should be taken to distinguish between beliefs and truths. There is no danger as long as a clear difference is made between temporary and proved explanations. For example, hypotheses and theories are attempts to explain natural phenomena. From these positions the scientist continues to experiment and observe until they are proved or discredited. The exact status of any explanation should be clearly labeled to avoid confusion.

The objectives of science are primarily the discovery and the subsequent understanding of the unknown. Man cannot be satisfied with recognizing that secrets exist in nature or that questions are unanswerable; he must solve them. Toward that end, specialists in the field of biology and related fields of interest are directing much of their time and energy.

Actually, two basic approaches lead to the discovery of new information. One, aimed at satisfying curiosity, is referred to as pure science. The other is aimed at using knowledge for specific purposes—for instance, improving health, raising standards of living, or creating new consumer products. In this case, knowledge is put to economic use. Such an approach is referred to as applied science.

Sometimes practical-minded people miss the point of pure science in thinking only of its immediate application for economic rewards. Chemists responsible for many of the discoveries could hardly have anticipated that their findings would one day result in applications of such a practical nature as those directly related to life and death. The discovery of one bit of information opens the door to the discovery of another. Some discoveries seem so simple that one is amazed they were not made years ago; however, one should remember that the construction of the microscope had to precede the discovery of the cell. The host of scientists dedicating their lives to pure science are not apologetic about ignoring the practical side of their discoveries; they know from experience that most knowledge is eventually applied.

(1) Pure science, leading to the construction of microscope, _____.

   A) may lead to antiscientific, "impure" results

   B) necessarily precedes applied science, leading to the discovery of a cell

   C) is not always as pure as we suppose

   D) necessarily results from applied science and the discovery of a cell

(2) A scientist interested in adding to our general knowledge about oxygen would probably call his approach _____.

   A) applied science   B) agriculture science

   C) pure science   D) environmental science

(3) Which of the following statements does the author imply?

   A) Scientists engaged in theoretical research should not be blamed for ignoring the practical side of their discoveries.

   B) Today few people have any notions of the meaning of science.

   C) In science, it is not difficult to distinguish fact from fiction.

   D) Practical-minded people can understand the meaning and objectives of pure science.

(4) The best title for the passage is _____.

   A) The Nature of Science and Scientists

   B) Biology and the Scientific Age

   C) Hypotheses and Theories

   D) On Distinguishing Fact from Fiction

Part B Reading

Ancient Technology

During the growth of the ancient civilizations, ancient technology was the result from advances in engineering in ancient times. These advances in the history of technology stimulated societies to adopt new ways of living and governance.

It was the growth of the ancient civilizations that produced the greatest advances in technology and engineering. This article includes the advances in technology and the development of several engineering arts before the Middle Ages.

Mesopotamia

Mesopotamian peoples (Sumerians, Akkadians, Assyrians and Babylonians) invented many technologies, most notably the wheel, which some consider the most important mechanical invention in history. Other Mesopotamian inventions include metalworking, copper-working, glassmaking, lamp making, textile weaving, flood control, water storage, as well as irrigation.

They were also one of the first Bronze Age people in the world. Early on they used copper, bronze and gold, and later they used iron. Palaces were decorated with hundreds of kilograms of these very expensive metals. Also, copper, bronze, and iron were used for armor as well as for different weapons such as swords, daggers, spears, and maces.

One of the first big ancient inventions was the invention of the wheel in eastern Mesopotamia. This was a very big "hit" in Mesopotamia because the wheel could transport goods like metal through the hilly geographical features and was more versatile and helped the need for agricultural transportation for food.

Egypt

The Egyptians invented and used many simple machines, such as the ramp to aid construction processes. They were among the first to extract gold by large-scale mining using fire-setting, and the first recognizable map, the Turin papyrus shows the plan of one such mine in Nubia.

Egyptian paper, made from papyrus, and pottery was mass produced and exported throughout the Mediterranean basin. The wheel, however, did not arrive until foreign invaders introduced the chariot. They developed Mediterranean maritime technology including ships and lighthouses.

Africa

Science and technology in Africa has a history stretching to the beginning of the human species, stretching back to the first evidence of tool use by hominid ancestors in the areas of Africa where humans are believed to have evolved. Africa saw the advent of some the earliest ironworking technology in the Air Mountains region of what is today Niger and the erection of some of the world's oldest monuments, pyramids and towers in Egypt, Nubia, and North Africa. In Nubia[1] and ancient Kush, glazed quartzite and building in brick was developed to a greater extent than in Egypt. Parts of the East African Swahili Coast saw the creation of the world's oldest carbon steel creation with high-temperature blast furnaces created by the Haya people of Tanzania.

Indian subcontinent

The Indus Valley Civilization, situated in a resource-rich area, is notable for its early application of city planning and sanitation technologies. Cities in the Indus Valley offer some of the first examples of closed gutters, public baths, and communal granaries. The Takshashila University was an important seat of learning in the ancient world. It was the center of education for scholars from all over Asia. Many Greek, Persian and Chinese students studied here under great scholars including Kautilya, Panini, Jivaka, and Vishnu Sharma.

Ancient India was also at the forefront of seafaring technology—

a panel found at Mohenjodaro, depicts a sailing craft. Ship construction is vividly described in the Yukti Kalpa Taru, an ancient Indian text on Shipbuilding.

Indian construction and architecture, called "Vaastu Shastra", suggests a thorough understanding of materials engineering, hydrology, and sanitation. Ancient Indian culture was also pioneering in its use of vegetable dyes, cultivating plants including indigo and cinnabar. Many of the dyes were used in art and sculpture. The use of perfumes demonstrates some knowledge of chemistry, particularly distillation and purification processes.

China

The Chinese were responsible for numerous technology discoveries and developments. Major technological contributions from China include early seismological detectors, matches, paper, the double-action piston pump, cast iron, the iron plough, the multi-tube seed drill, the suspension bridge, natural gas as fuel, the magnetic compass, the raised-relief map, the propeller, the crossbow, the South Pointing Chariot, and gun powder.

Persia

The Qanat[2], a water management system used for irrigation, originated in Iran before the Achaemenid period of Persia. The oldest and largest known qanat is in the Iranian city of Gonabad which, after 2700 years, still provides drinking and agricultural water for nearly 40,000 people.

Persian philosophers and inventors may have created the first batteries, sometimes known as the Baghdad Battery, in the Parthian or Sassanid eras. Some have suggested that the batteries may have been used medicinally. Other scientists believe the batteries were used for electroplating—transferring a thin layer of metal to another metal surface—a technique still used today and the focus of a common classroom experiment. In the 7th century A.D., Persians in Afghanistan developed the first practical windmills.

Greek and Hellenistic[3]

Greek and Hellenistic engineers invented many technologies and improved upon pre-existing technologies, particularly during the Hellenistic period. Hero of Alexandria invented a basic steam engine and demonstrated knowledge of mechanic and pneumatic systems. Archimedes[4] invented several machines. The Greeks were unique in pre-industrial times in their ability to combine scientific research with the development of new technologies. One example is the Archimedean screw; this technology was first conceptualized in mathematics, and then built. Other technologies invented by Greek scientists include the ballistae, the piston pump, and primitive analog computers like the Antikythera mechanism. Greek architects were responsible for the first true domes, and were the first to explore the Golden ratio and its relationship with geometry and architecture.

Apart from Hero of Alexandria's steam aeolipile; Hellenistic technicians were the first to invent watermills and wind-wheels, making them global pioneers in three of the four known means of non-human propulsion prior to the Industrial Revolution. However, only water power was used extensively in antiquity.

Other Greek inventions include torsion catapults, pneumatic catapults, crossbows, cranes, runways, organs, the keyboard mechanism, gears, differential gears, screws, refined parchment, showers, dry docks, diving bells, odometer and astrolabes. In architecture, Greek engineers constructed monumental lighthouses such as the Pharos and devised the first central heating systems. The Tunnel of Eupalinos is the earliest tunnel in history which has been excavated with a scientific approach from both ends.

Automata like vending machines, automatic doors and many other ingenious devices were first built by Hellenistic engineers as Ctesibius[5], Philo of Byzantium[6] and Heron. Greek technological treatises were scrupulously studied and copied by later Byzantine, Arabic and Latin European scholars and provided much of the foundation for further technological advances in these civilizations.

Unit Two   Ancient Civilizations

Roman

Romans developed an intensive and sophisticated agriculture, expanded upon existing iron working technology, created laws providing for individual ownership, advanced stonemasonry technology, advanced road-building (exceeded only in the 19th century), military engineering, civil engineering, spinning and weaving and several different machines like the Gallic reaper that helped to increase productivity in many sectors of the Roman economy. They also developed water power through building aqueducts on a grand scale, using water not just for drinking supplies but also for irrigation, powering water mills and in mining. They used drainage wheels extensively in deep underground mines; one device being the reverse overshot water-wheel. They were the first to apply hydraulic mining methods for prospecting for metal ores, and for extracting those ores from the ground when found using a method known as hushing.

Roman engineers build monumental arches, amphitheatres, aqueducts, public baths, true arch bridges, harbours, dams, vaults and domes on a very large scale across their Empire. Notable Roman inventions include the book (*Codex*), glass blowing and concrete. Because Rome was located on a volcanic peninsula, with sand which contained suitable crystalline grains, the concrete which the Romans formulated was especially durable. Some of their buildings have lasted 2000 years, to the present day.

Roman civilization was highly urbanized by pre-modern standards. Many cities of the Roman Empire had over 100,000 inhabitants with the capital Rome being the largest metropolis of antiquity. Features of Roman urban life included multistory apartment buildings called insulae, street paving, public flush toilets, glass windows and floor and wall heating. The Romans understood hydraulics and constructed fountains and waterworks, particularly aqueducts, which were the hallmark of their civilization. They exploited water power by building water mills, sometimes in series, such as the sequence found at Barbegal in southern France and suspected on the Janiculum in Rome. Some Roman baths have lasted to this day. The

Romans developed many technologies which were apparently lost in the Middle Ages, and were only fully reinvented in the 19th and 20th centuries. They also left texts describing their achievements, especially Pliny the Elder, Frontinus and Vitruvius.

Other less known Roman innovations includes cement, boat mills, arch dams and possibly tides mills.

New Words and Expressions

Mesopotamia 美索不达米亚
Akkadian 阿卡德人
Assyrian 亚述人
Babylonian 巴比伦人
textile 纺织的
armor 盔甲
dagger 短剑，匕首
mace 锤矛（中世纪作武器用）
ramp 斜面；斜坡；坡道
papyrus 纸莎草
chariot 双轮战车
Niger 尼日尔（西非国家）
quartzite 石英岩
Swahili Coast 斯瓦希里海岸
Tanzania 坦桑尼亚（东非国家）
sanitation 公共卫生，环境卫生
gutter （道路的）排水边沟，街沟
communal 公有的，共有的
granary 谷仓，粮仓
seafaring 航海的；以航海为业的
hydrology 水文学

indigo 槐蓝属植物
cinnabar 朱砂
distillation 蒸馏；净化
seismological 地震学上的
piston pump 活塞泵
cast iron 铸铁；生铁
seed drill 谷物条播机
magnetic compass 磁罗盘
crossbow 石弓；弩
Parthian 帕提亚的
Sassanid 萨桑君王的
electroplate 电镀
Afghanistan 阿富汗
pneumatic 压缩空气推动的，气动的
conceptualize 概念化
ballista [复数 ballistae] 弩炮
dome 圆屋顶，穹顶
aeolipile 汽转球
propulsion 推进（力）
torsion 扭力；扭转；转矩
catapult （古代的）石弩；投石机

Unit Two　Ancient Civilizations

> crane 起重机，吊车
> gear 齿轮
> parchment 羊皮纸；仿羊皮纸
> diving bell 潜水钟
> odometer 里程计
> tunnel 隧道，地道
> excavate 挖掘（穴、洞等）；开凿
> automata 自动操作；自动化（技术）
> ingenious 制作精巧的
> treatise （专题）论文；专著
> scrupulously 细致的；一丝不苟的
> stonemasonry 石砌体
> reaper 收割机
> aqueduct 沟渠；导水管
> drainage 排水
> hydraulic 水力的
> hush 用水清洗（矿石）
> amphitheatre 圆形露天剧场
> vault 拱顶
> codex （圣经、经典著作等的）手抄本
> urbanize 使都市化
> metropolis 首都，首府；大都会
> multistory 多层建筑
> insulae [insula 的复数] 脑岛
> flush 冲洗

Notes

1. Nubia 努比亚。指埃及尼罗河第一瀑布阿斯旺与苏丹第四瀑布库赖迈之间的地区。

2. Qanat（伊朗等国从山上引水至平原的）暗渠，坎儿井。

3. Hellenistic（公元前 4 世纪至公元前 1 世纪的）希腊时期的。

4. Archimedes（公元前 287—前 212）阿基米德，古希腊哲学家、数学家、物理学家。

5. Ctesibius 特西比乌斯，希腊发明家。

6. Philo of Byzantium 拜占廷人斐罗。他写下《世界七大奇迹》，其中使用 Pyramid of Giza（中译"吉萨金字塔"，释意为"在吉萨的锥体"）来命名法老卡夫拉的陵墓群。

Exercises

I. Find out the English equivalents to the following Chinese terms from the passage.

1. 机械发明
2. 金属加工
3. 玻璃吹制
4. 蓄水
5. 青铜器时代
6. 多山的地理特征
7. 碳钢
8. 高炉
9. 早期地震探测仪器
10. 吊桥
11. 模拟计算机
12. 黄金比例
13. 中央供暖系统
14. 自动贩卖机
15. 水力采矿
16. 文明的标志

II. Answer the following questions by using the sentences in the passage or in your own words.

1. What is considered to be one of the most important mechanical inventions in history by Mesopotamian peoples? Why?
2. What is Vaastu Shastra?
3. Give an example to show why the Greeks were unique in pre-industrial times to combine scientific research with the development of new technologies. List some technologies invented by Greek architects and engineers.
4. What are the advances made by the ancient Romans?

Part C Extended Reading

Development of Civilization

The first two stages in the development of civilized man were probably the invention of primitive weapons and discovery of fire, although nobody knows when he acquired the use of the latter.

The origin of language is also obscure. No doubt it began very gradually. Animals have a few cries that serve as signals, but even the highest apes have not been able to pronounce words, even with the most intensive professional instruction. The superior brain of man is apparently a necessity for the mastering of speech. When man became sufficiently intelligent, we must suppose that he gradually increased the number of cries for different purpose. It was a great day when he discovered that speech could be used for narrative. There are those who think that in this respect picture language preceded oral language. A man could draw a picture on the wall of his cave to show in which direction he has gone, or what prey he hoped to catch. Probably picture language and oral language developed side by side. I am inclined to think that language has been the most important single factor in the development of man.

Two important stages came not so long before the dawn of written history. The first was the domestication of animals; the second was agriculture. Agriculture was a step in human progress to which subsequently there was nothing comparable until our own machine age. Agriculture made possible an immense increase in the number of the human species in the regions where it could be successfully practiced. These were, at first, only those in which nature fertilized the soil after each harvest. Agriculture met with violent resistance from the pastoral nomads, but the agricultural way of life prevailed in the end because of the physical comforts it provided.

Another fundamental technical advance was writing, which, like spoken language, developed out of pictures, but as soon as it had reached a certain stage, it was possible to keep records and transmit information to people who were not present when the information was given.

These inventions and discoveries—fire, speech, weapons, domestic animals, agriculture, and writing—made the existence of civilized communities possible. From about 3000 B.C. until the beginning of the Industrial Revolution less than two hundred years ago, there was no technical advance comparable to these. During this long period man had enough time to become accustomed to his technique, and to develop the beliefs and political organizations appropriate to it. There was, of course, an immense extension in the area of civilized life. At the end of this long period, it covered much the greater part of the inhabitable globe. I do not mean to suggest that there was no technical progress during this long time. There was progress—there were even two inventions of very great importance, namely gunpowder and the mariner's compass—but neither of these can be compared in their revolutionary power to such things as speech and writing and agriculture.

New Words and Expressions

obscure 模糊的，晦涩的
intelligent 智慧的，开化的
narrative 描述性的，叙述的
precede 先于……
prey 被捕食的动物，捕获物
be inclined to 倾向于……的
dawn 开端
subsequently 随后，后来，接着
resistance 阻力
prevail 流行，盛行，普遍

Unit Two   Ancient Civilizations

📝 Exercises

I. Find out the English equivalents to the following Chinese terms from the passage.

1. 原始武器
2. 最专门的强化训练
3. 绘画语言
4. 口头语言
5. 同步发展
6. 文字记载的历史开端
7. 动物的驯养
8. 机器时代
9. 游牧部落成员
10. 农业生活方式
11. 舒适的物质生活
12. 文明社会的存在

II. Translate the following sentences into Chinese.

1. When man became sufficiently intelligent, we must suppose that he gradually increased the number of cries for different purpose. It was a great day when he discovered that speech could be used for narrative. There are those who think that in this respect picture language preceded oral language.

2. Two important stages came not so long before the dawn of written history. The first was the domestication of animals; the second was agriculture. Agriculture was a step in human progress to which subsequently there was nothing comparable until our own machine age.

3. Agriculture met with violent resistance from the pastoral nomads, but the agricultural way of life prevailed in the end because of the physical comforts it provided.

4. These inventions and discoveries—fire, speech, weapons, domestic animals, agriculture, and writing—made the existence of civilized communities possible. From about 3000 B.C. until the beginning of the Industrial Revolution less than two hundred years ago there was no technical advance comparable to these.

5. I do not mean to suggest that there was no technical progress during this long time. There was progress—there were even two inventions of very

great importance, namely gunpowder and the mariner's compass—but neither of these can be compared in their revolutionary power to such things as speech and writing and agriculture.

Unit Three

History of Science in Early Cultures

Part A Lecture

掌握科技英语文献的阅读方法

当今时代是信息时代，人们要能迅速无误地从一段文字中最大限量地获取信息，需要有较强的阅读能力。阅读能力强的读者能够根据不同的阅读目的，选取适当的阅读方法进行有效的阅读，提高阅读效率。阅读的目的不同，阅读材料的难度不同，阅读的方法和技巧也不同。根据阅读目的和材料难度，科技英语文献常用的阅读方法有快速阅读（fast reading）和仔细阅读（简称细读）（reading for full understanding）两种。快速阅读是一种既要求理解度又要求速度的阅读方法，包括略读（skimming）和查读（scanning）。略读是为了快速获取信息；查读是为了快速获取数据。仔细阅读的目的是为了领会和掌握文献的信息。科技文献的很多内容是浓缩的，里面有大量的信息，要理解和掌握这些内容，需采用细读的方法。本讲座将详述快速阅读和仔细阅读的技巧。

1 快速阅读技巧

快速阅读是以较快的速度在规定的时间内有目的、有方法、高效率地阅读材料，以便从中获得所需信息的阅读方法。在文字信息迅猛增长的现代信息社会，培养略读和查读这两种基本快速阅读技能可以使学生极大地提高阅读的速度和效率，在短时间内大量、快速地获取文字信息，捕捉、提取主要信息，迅速做出正确的概括和判断。

1·1 略读

略读又称跳读（reading and skipping）或浏览（glancing），是一种非常实用的快速阅读技能。所谓略读，是指以尽可能快的速度跳跃式地阅读文

Unit Three　History of Science in Early Cultures

献以获取文章大意或中心思想的阅读方法。略读要求读者在不细读全文的前提下了解文章的大意，初步形成对文章的主观印象（mental image），为进一步阅读做好准备。

略读的目的在于尽快了解整篇文章的梗概，所以略读时需要把注意力集中在寻找读物的中心思想而不是细节上；做到不逐词阅读而只阅读关键词，如名词、动词等实词和表示句际和段际关系的连接词；选择重点部分阅读，如题目、小标题、粗体字、斜体字、划线部分、重点句子（主题句）、重点段落、开头和结尾部分等。在平常的阅读实践中，我们通过"略读"的技能来识别文章的类型，找到文章主题，识别作者的观点，并发现一些与中心思想紧密关联的事实等。

下面让我们通过实例了解如何运用略读技能查找主题句、中心思想和主要信息等。

Directions: *Read the following paragraph and find the topic sentence.*

　　The Florida landscape boasts a wide variety of plant life. Almost half of all the different kinds of trees found in America grow in Florida. Some of Florida's woodlands are filled with majestic pine trees. Swamp maples, bald cypresses, bays, and oaks grow in some of the state's forests. Dozens of different kinds of subtropical trees can be found in the Florida peninsula and the Keys. The warm climate in these areas nourishes the strangler fig, royal palm, and mangroves, for example.

分析：该段落的第1句为主题句，后面的句子是具体介绍各种植物。在略读中，需要迅速判断文章的关键词和关键句子，这有助于确定文章中哪些是主要信息，哪些是次要信息。

Directions: *Read the following paragraph and determine the main idea.*

　　Someday we will all have robots that will be our personal servants. They will look and behave much like real humans. We will be able to talk to these mechanical helpers and they will be able to respond in kind. Amazingly, the robots of the future will be able to learn from experience. They will be smart, strong, and untiring workers whose only goal will be to make our lives easier.

Which sentence from the paragraph expresses the main idea?

A) Someday we will all have robots that will be our personal servants.

B) We will be able to talk to these mechanical helpers and they will be able to respond in kind.

C) They will look and behave much like real humans.

D) Amazingly, the robots of the future will be able to learn from experience.

分析：从段落中可知，A 项为主题句，阐述了本段的中心思想，其余句子都是用来支撑主题句的。

1. Directions: *Skim the following passage very quickly. Take no more than 20 seconds. And then answer the following questions.*

Railway Network

Britain's railways are big business. They carry more than two million passengers a day on about 16,000 trains. They also move 125 million tons of freight every year. British Rail's 23,500 miles of track, more than a quarter of it electrified, is very busy.

Railways have been part of the British heritage for more than 150 years. At first, people did not understand how dangerous the railways could be, but important lessons were learned every time an accident happened. That is why accidents are now so rare.

Road accidents claim more lives every month than the total number of passengers killed in train accidents on British Rail in the last ten years.

Far more people are killed or injured while trespassing on the railway. The toll may not be very high compared with road casualties but the grief, pain and the suffering of relatives are just as terrible.

(1) What is the passage about?

(2) Does it give information?

(3) Does it give an opinion?

(4) Is it fact or fiction?

Unit Three   History of Science in Early Cultures

(5) Is it aimed at adults or children?

分析：略读文章后很快可以找到答案：(1) Railways; (2) Yes; (3) No; (4) Fact;
(5) Mainly adults, but older children could understand it too.

2. Directions: *Now read the passage again at a slightly slower speed. Take no more than 30 seconds. See how much more information you have taken in by answering the following questions.*

(1) What else besides passengers do the trains carry?
(2) For how long have railways been part of the British heritage?
(3) Which claims more lives, railways or roads?
(4) How are more people killed or injured on the railways?

分析：从文章中可以很快找到答案：(1) Freight; (2) More than 150 years;
(3) Roads; (4) By trespassing on the railways。

从以上例子可看出，略读的关键是"略"。略读的目的是为了对想阅读的文章进行一般、初步的了解，就是了解作者的主要观点和文章的中心思想。略读中只要求抓住文章的大概内容，初步摸清文章的脉络，把注意力放在与全文主题有关的信息上。一般说来，每段文字中都有一个主题句或暗藏着主题句，如找出各段的主题句并把它们集中起来加以总结，就可得出全文的中心思想。

略读抓大意的能力在实际阅读中的用途越来越广。人们的时间有限，精力有限，而要读的东西又多。在这种情况下能通过略读抓住要点或大意就可以掌握最基本的信息，解决阅读的实际问题。

1·2 查读

查读又称寻读，也是一种快速阅读技巧。查读是一种从大量的资料中迅速查找某一项具体事实或某一项特定信息，如人物、事件、时间、地点、数字等，而对其他无关部分则略去不读的快速阅读方法。运用这种方法，读者就能在最短的时间内略读尽可能多的印刷材料，找到所需要的信息。因此，查读既要求速度，又要求准确性。

为了有效地进行查读，我们可以运用以下技巧：1）利用材料的编排形式。不管是什么资料，它们都是按照某种逻辑方法排列的。有的按字母顺序排列，如词典、索引、邮政编码簿、电话号码簿以及其他参考资料；有的按时间顺序排列，如电视节目、历史资料等；有的按类别排列，如报纸上的体育版面等。又如，新闻报道往往先传达主要信息，然后遵循事件发生的时间地点、发展和原因等逻辑顺序编排。2）利用文章标题和副标题。查读时，先看文章标题或章节标题，确定文章哪一部分包含自己所需要的材料，这样可以直接翻到那个部分，进行查找。3）抓提示词。查读时，找到包含所需信息的章节，然后迅速扫描，要特别留意相关信息的提示词。找到提示词，就可以采用一般阅读速度，获得所需要的信息。

Directions: *Scan the passage quickly. And answer the following questions.*

The climate of any place is the kind of weather it usually has over a long period of time. The kinds of homes we live in, the clothes we wear, and even the foods we eat depend on the climate of the place where we live.

Climate is a complicated pattern; it is affected by many things. Nearness to the North or South Pole or to the equator is important. If you live near one of the poles, you live in a cold climate, for you do not get as much or as direct sunshine as you would get farther from the poles. If you live near the equator, you live in a warm or very hot climate, for this is the region where the sun shines almost straight down.

How much rain or snow falls makes a great difference to the climate. You may live in a hot, dry land, where little rain falls. This will be a desert. Its climate is quite different from that of a rain-forest, which may be the same distance from the equator but where rain falls almost every day. The amount of rain that falls—or snow, in a cold land—depends upon the currents in nearby seas. Rainfall depends on many different things.

(1) Which is the general weather of an area over a long period of time called?

(2) Name one important thing on which climate depends.

(3) Name one important thing on which rainfall depends.

(4) What is a hot, dry land where little rain falls?

Unit Three　History of Science in Early Cultures

(5) If you live near the North or South Pole, what kind of climate do you have?

分析：

第 1 题要求我们给出"一个地区长时期天气情况"的简称（或定义）。答案就在第一段首句，即：climate（气候）。

第 2 题要求找出气候所依赖的重要因素。第二段第 2 句是关键句，它表明与地球两极或地球赤道距离的远近是影响气候的重要因素。另外，文章第三段的首句也表明，影响气候的因素还有雨量或降雪量，因此，该题答案是 nearness to poles or equator, and amount of rain or snow。

第 3 题要找出降雨所依赖的因素。文章第三段的倒数第 2 句是关键句。该句提到三点：the winds; the surrounding mountains; the sea currents。正确答案便是其中之一。

第 4 题要求查找"炎热、干旱以及降雨量少的地方"。作者在第三段的第 2、3 句中讲到这条信息：This will be a desert。因此，答案是 desert。

第 5 题问的是假如生活在北极或南极，会遇到什么样的气候？对于这道题，可迅速找到作者在文章什么地方提及北极和南极，这是最有可能找到答案的地方。此题的答案应是 cold。从本例可知，对某一主题的背景知识知道的越多，查找信息的速度就越快。因此，个人知识的多与少会直接影响到阅读某些文章的速度。

本例也告诉我们，查读的关键是"目标"。查读要有查找目标，以较快的速度从大量材料中捕捉有关信息。查读时，尽快扫视所读材料，一旦发现有关的内容，就要将它记住或摘下，既保证查读的速度，又做到准确无误。找到所需的信息，阅读便可终止，不必读完全文。熟练的读者善于运用查读获得具体信息，以提高阅读效率。

1·3　略读和查读的区别

略读和查读不同。略读的目的是为了抓住文章的大意和主要内容，而查读的目的则是为了获取文章中特定的信息。略读前，读者没有明确的目的，对读物完全不了解或了解很少；而查读之前，读者带有较为明确的目的性，对读物的内容有了初步的了解，并且还知道查找什么，在哪里查，以及如何查。查读的速度较之略读更快。

2 仔细阅读技巧

仔细阅读是一种边阅读边理解边深入思考的阅读方法，是必须掌握的高级阅读技能。仔细阅读时要求对文章逐段逐句仔细阅读，辨认句子之间的逻辑关系，深入思考重要的词、句，以求得较深刻准确的理解，既能弄清文章字面的意思，也能理解"字里行间"的含义，达到融会贯通。仔细阅读不等于慢读，效率高的读者需要区分相关细节，理解彼此间的关系，以及段落、章节的中心思想之间的关系，并对相关细节进行同化；注意集中领会文献的信息，立足于读一遍就能理解全部内容。阅读文体不熟悉、专业词汇较多的英语科技文献需要放慢速度，仔细阅读。细读是为了领会和掌握文献的信息，提高英语理解水平。

仔细阅读是学好英语的关键。仔细阅读的方法主要包括预览、结构通览、列提纲、笔录、划线和总结等。

2.1 预览

高效的读者会把预览（previewing）所要阅读材料的内容作为阅读的重要一步。获取所要阅读材料的大致情况，可以有效增进理解和加强记忆力。

2.2 结构通览

结构通览（structured overview）是通过抓住关键词语来达到看清整篇文章的轮廓以及各概念之间关系的目的。结构通览一般需要10分钟左右时间，由文章的长短、难易度而定。

2.3 列提纲

提纲是文章的骨架，勾画了文章的轮廓。列提纲（outlining）是厘清思路、规划文章结构的重要过程。

提纲一般分为两大类：标题式提纲（topic outline）和主题式提纲（sentence

Unit Three　History of Science in Early Cultures

outline）。用短语和词组列出文章层次、文章段落、文章各部分的顺序，即为标题式提纲。用完整的句子列出来的提纲叫主题式提纲。这些完整的句子实际上就是成文之后的各段主题句。如：

　　Erosion is the transporting of weathered materials downhill. It may occur in two ways. One way is through an agent of transportation. A second way is through gravity in the form of landslides or slow creep.

　　Agents of transportation vary. Streams are a primary agent. They transport large amounts of sand, gravel, and boulders. When the stream is the agent, it deposits its fragments in different places according to their sizes. The largest boulders are deposited in the upper reaches of the stream. Medium particles such as gravel will be carried further downstream. Silt and clay will travel the furthest; they may move out of the stream when it floods, or even be carried out to sea. The wind is a second major agent of erosion. In deserts the wind carries along sand, gravel, silt, and clay. Ice is the third agent.

分析：列提纲是一项复杂的技能，因为它不仅要识别中心思想，还需把相关的细节与中心思想联系起来以显示出逻辑的程序框架。本篇的标题式提纲和主题式提纲见表 3-1。

表 3-1　标题式提纲和主题式提纲

标题式提纲	主题式提纲
(Title) Occurrence of Erosion	(Title) Erosion Occurs in Two Ways
I. Through an agent of transportation	I. It is through an agent of transportation
A. Streams	A. Streams are a primary agent
1. The largest boulders	1. The largest boulders are deposited in the upper reaches of the streams.
2. Medium particles	2. Medium particles are carried further downstream.
3. Silt and clay	3. Silt and clay travel furthest and they may move out of the stream or be carried out to sea.
B. Wind	B. The wind is the second major agent.
C. Ice	C. Ice is the third agent.
II. Through gravity	II. It is through gravity in the form of landslides or slow creep

2.4 笔录

有效的笔录（note-taking）是一种高水平的组织技能。它和列提纲不同，主要表现在笔录是别人讲，自己写，自己看。提纲一方面是自己写好后，再铺开来写成文章，使其呈现整体感；另一方面是勾画出阅读文章的轮廓，以便吸取其精华。笔录和列提纲也有共性，即需要区别哪些是主要信息，哪些是次要信息；识别各种信息间的关系。养成良好的笔录习惯，可以提高我们组织信息的能力。

2.5 划线

划线（underlining）是一种常见的阅读方法，划线需要学生一步一步来完成以下步骤：1）先概观一下整个章节；2）再逐段逐节地阅读；3）在读了一段、看了一节之后，要划出中心思想和铺陈细节的关键词和短语。可以用彩色笔，一种颜色划中心句，一种颜色划铺陈句。划线能使我们快速方便地复习所学内容，更有利于抓住重点。

2.6 总结

总结（summarizing）是指用简单的形式阐述中心思想以及重要事例，它是帮助学生记住所学材料的一种行之有效的方法。看完一个章节，回过头来想一想这一章节主要讲什么，这是总结；看完一本小说，写一篇书评，这也是总结；平时上课之后，回想一下上课的内容，结合课堂上记录的内容，写些心得，这也是总结。

写总结有利于全面理解和掌握文章的内容要点，并且有利于提高分析问题、综合问题的能力和语言表达能力。写总结时必须善于抓住原文的主要情节或中心内容，勇于舍弃那些辅助作用的细枝末节。

Direction: *Read the passage carefully and write a summary.*

Smoking and Cancer

Americans smoke six thousand million cigarettes every year (1970

Unit Three   History of Science in Early Cultures

figures). This is roughly the equivalent of 4195 cigarettes a year for every person in the country of 18 years of age or more. It is estimated that 51% of American men smoke compared with 34% of American women.

Since 1939, numerous scientific studies have been conducted to determine whether smoking is a health hazard. The trend of the evidence has been consistent and indicates that there is a serious health risk. Research teams have conducted studies that show beyond all reasonable doubt that tobacco smoking, particularly cigarette smoking is associated with a shortened life expectancy.

Cigarette smoking is believed by most research workers in this field to be an important factor in the development of cancer of the lungs and cancer of the throat and is believed to be related to cancer of the bladder and the oral cavity. Male cigarette smokers have a higher death rate from heart disease than non-smoking males. (Female smokers are thought to be less affected because they do not breathe in the smoke so deeply.) The majority of physicians and researchers consider these relationships proved to their satisfaction and say, "Give up smoking. If you don't smoke—don't start!"

Some competent physicians and research workers—though their small number is dwindling even further—are less sure of the effect of cigarette smoking on health. They consider the increase in respiratory diseases and various forms of cancer may possibly be explained by other factors in the complex human environment—atmospheric pollution, increased nervous stress, chemical substances in processed food, or chemical pesticides that are now being used by farmers in vast quantities to destroy insects and small animals. Smokers who develop cancer or lung diseases, they say, may also, by coincidence, live in industrial areas, or eat more canned foods. Gradually, however, research is isolating all other possible factors and proving them to be statistically irrelevant.

Apart from statistics, it might be helpful to look at what smoking tobacco actually does to the human body. Smoke is a mixture of gases, vaporized chemicals, minute particles of ash, and other solids. There is also nicotine, which is a powerful poison, and black tar. As the smoke is breathed in, all these components form deposits on the membranes of the lungs. One point of concentration is where the air tube, or bronchus, divides. Most lung cancer begins at this point.

Smoking also affects the heart and blood vessels. It is known to be related to Beurger's disease, a narrowing of the small veins in the hands and feet that can cause great pain and lead even to amputation of limbs. Smokers also die much more often from heart disease.

While all tobacco smoking affects life expectancy and health, cigarette smoking appears to have a much greater effect than cigar or pipe smoking. However, nicotine consumption is not diminished by the latter forms, and current research indicates a causal relationship between all forms of smoking and cancer of the mouth and throat. Filters and low tar tobacco are claimed to make smoking to some extent safer, but they can only marginally reduce, not eliminate the hazards.

分析：写总结时首先要准确理解原文，搞清原文中所叙事件之间、人与人之间、观点与观点之间的关系，搞清原文中的要点与次要内容，然后抓住主要情节或主要论点组织成篇。写总结的注意事项：1）写成的总结必须是一个有机的整体，较全面地概括原文；2）要忠实于原文，不能遗漏原文中主要论点。根据短文内容，本篇的总结如下：

This passage deals with smoking and cancer. Nearly half the adult population of the U. S. A. are smokers, in spite of the fact that medical research has shown beyond reasonable doubt that smoking is associated with poor health. One very important disease with which smoking is thought to be connected, is cancer. However, a few doctors think such things as bronchitis and cancer have other causes, but their view is not well-supported statistically. Although all forms of smoking are almost certainly dangerous, cigarettes appear to be the most harmful.

3 综合练习

本节将分析快速阅读和仔细阅读技巧在科技文章中的综合应用情况。

1. **Directions:** *Skim the following passage, and then choose the correct answer to each question.*

Unit Three   History of Science in Early Cultures

Pulp Friction

[1] Every second, 1 hectare of the world's rain-forest is destroyed. That's equivalent to two football fields. An area the size of New York City is lost every day. In a year, that adds up to 31 million hectares—more than the land area of Poland. This alarming rate of destruction has serious consequences for the environment; scientists estimate, for example, that 137 species of plant, insect or animal become extinct every day due to logging. In British Columbia, where, since 1990, thirteen rain-forest valleys have been clear-cut, 142 species of salmon have already become extinct, and the habitats of grizzly bears, wolves and many other creatures are threatened. Logging, however, provides jobs, profits, taxes for the government and cheap products of all kinds for consumers, so the government is reluctant to restrict or control it.

[2] Much of Canada's forestry production goes towards making pulp and paper. According to the Canadian Pulp and Paper Association, Canada supplies 34% of the world's wood pulp and 49% of its newsprint paper. If these paper products could be produced in some other way, Canadian forests could be preserved. Recently, a possible alternative way of producing paper has been suggested by agriculturalists and environmentalists: a plant called hemp.

[3] Hemp has been cultivated by many cultures for thousands of years. It produces fiber which can be made into paper, fuel, oils, textiles, food, and rope. For centuries, it was essential to the economies of many countries because it was used to make the ropes and cables used on sailing ships; colonial expansion and the establishment of a world-wide trading network would not have been feasible without hemp. Nowadays, ships' cables are usually made from wire or synthetic fibers, but scientists are now suggesting that the cultivation of hemp should be revived for the production of paper and pulp. According to its proponents, four times as much paper can be produced from land using hemp rather than trees, and many environmentalists believe that the large-scale cultivation of hemp could reduce the pressure on Canada's forests.

[4] However, there is a problem: hemp is illegal in many countries of the world. This plant, so useful for fiber, rope, oil, fuel and textiles, is a species of cannabis, related to the plant from which marijuana is produced. In the late 1930s, a movement to ban the drug marijuana began to gather force,

resulting in the eventual banning of the cultivation not only of the plant used to produce the drug, but also of the commercial fiber-producing hemp plant. Although both George Washington and Thomas Jefferson grew hemp in large quantities on their own land, any American growing the plant today would soon find himself in prison—despite the fact that marijuana cannot be produced from the hemp plant, since it contains almost no THC (the active ingredient in the drug).

[5] In recent years, two major movements for legalization have been gathering strength. One group of activists believes that ALL cannabis should be legal—both the hemp plant and the marijuana plant—and that the use of the drug marijuana should not be an offense. They argue that marijuana is not dangerous or addictive, and that it is used by large numbers of people who are not criminals but productive members of society. They also point out that marijuana is less toxic than alcohol or tobacco. The other legalization movement is concerned only with the hemp plant used to produce fiber; this group wants to make it legal to cultivate the plant and sell the fiber for paper and pulp production. This second group has had a major triumph recently: in 1997, Canada legalized the farming of hemp for fiber. For the first time since 1938, hundreds of farmers are planting this crop, and soon we can expect to see pulp and paper produced from this new source.

(1) The main idea of Paragraph One is that _____.
    A) scientists are worried about New York City
    B) logging is destroying the rain-forests
    C) governments make money from logging
    D) salmon is an endangered species

(2) The main idea of Paragraph Two is that _____.
    A) Canadian forests are especially under threat
    B) hemp is a kind of plant
    C) Canada is a major supplier of paper and pulp
    D) Canada produces a lot of hemp

(3) The main idea of Paragraph Three is that _____.
    A) paper could be made from hemp instead of trees

Unit Three History of Science in Early Cultures

   B) hemp is useful for fuel

   C) hemp has been cultivated throughout history

   D) hemp is essential for building large ships

(4) The main idea of Paragraph Four is that _____.

   A) hemp is used to produce drugs

   B) many famous people used to grow hemp

   C) it is illegal to grow hemp

   D) hemp is useful for producing many things

(5) The main idea of Paragraph Five is that _____.

   A) hemp should be illegal because it is dangerous

   B) many people have been working to legalize hemp recently

   C) hemp was made illegal in 1938

   D) marijuana is not a dangerous drug

分析：略读短文后可知，以上题目的答案分别是 B、C、A、C、B。

2. **Directions:** *Read the passage again, and scan the correct answer to each question.*

(1) How many species of salmon have become extinct in B.C.?

   A) 27   B) 31   C) 137   D) 142

(2) How much of the world's newsprint paper is supplied by Canada?

   A) 31%   B) 49%   C) 34%   D) 19%

(3) What equipment on a ship was made from hemp?

   A) Ropes   B) waterproof cloth   C) engine fuel   D) life rafts

(4) What drug can be obtained from a relative of hemp?

   A) Cocaine   B) Heroin   C) Amphetamine   D) Marijuana

(5) Where was hemp farming recently legalized?

   A) The USA   B) Canada   C) Singapore   D) The Netherlands

分析：通过查读短文可知，以上题目的答案分别是 D、B、A、D、B。

从上例可知，本题要求采用略读和查读的阅读方法解题。在对全文进行略读后，我们对文章的内容、范围、结构等有一个整体印象。再根据问题对文章查读，迅速扫描问题答案可能所在的地方，进行精读，以求得更深刻的理解。由此看出，采用略读和查读的方法可以使我们在做阅读理解题时大大提高准确性和速度。

Directions: *Skim or preview the main idea of every paragraph and then write the outlines of the passage.*

Straight and Crooked Thinking

[1] If we observe the actions of men, whether as individuals or as groups, and whether scientists or non-scientists, we find that they frequently fall into avoidable error because of a failure to reason correctly. There are many reasons for this, though only a few can be dealt with here.

[2] The first difficulty is bound up with the use of words. It frequently happens that what one person means when he uses a certain word is different from what others mean. Consider, for example, the word intelligence, oxygen, accurate and average. In intelligence we face the problem that a word may not mean only one thing, but many—in this instance a very complicated set of aptitudes and abilities whose number and characteristics are not agreed upon by the specialists who study the phenomenon, and are even less understood by the layman. In oxygen we have a different problem, for although both a research chemist and a chemical manufacturer identify the word theoretically with the element O, in practice they have different concepts about it. Thus if the researcher performed a delicate experiment, using the manufacturer's oxygen, it might easily be a failure since the so-called O, whether used as a solid, liquid or gas, would almost certainly contain other substances. Hence another difficulty about words is that they often do not differentiate clearly enough between several varieties of the "same" thing.

[3] Another common error connected with words consists in confusing a word or a name with a fact. The course of scientific progress has been frequently slowed down by 1) assuming the existence of something to account for a certain phenomenon, 2) giving the assumed substance a name, e.g. phlogiston, aether, etc., and 3) implying that the phenomenon

Unit Three History of Science in Early Cultures

has been satisfactorily accounted for.

[4] Apart from the misuse of words, mistakes in logic can occur. Thus an example is recorded of a young sociologist, investigating literacy in a certain community, who discovered from the official records that over 50 percent of the population was females. He subsequently found that approximately 70 per cent of the population was literate. When he had obtained this data he summed it up and drew conclusions as follow:

> Most of the population is females;
>
> Most of the population is literate;
>
> Most females are literate.

[5] This was, of course, an unreasonable inference, as the investigator himself realized as soon as he had re-examined his chain of reasoning more carefully.

[6] Another mistake is to confuse cause and effect. This may easily occur at the beginning of an investigation, but if it remains uncorrected it can be considered as primarily a by-product of insufficient experimentation. To illustrate this, the following case can be quoted. The inhabitants of a certain community had noted over the ages that whenever an individual became ill with a fever, the body parasites left him. They therefore made the correlation that the parasites kept them healthy. Later, however, properly-controlled scientific investigation showed that the reverse was true: in fact the parasites transmitted several kinds of fever and then left the sick persons when the latter's bodies became too hot to live on.

[7] Some other factors which may influence reasoning are 1) faulty analogizing, 2) the inhibiting effect on further research of concepts which have been widely-accepted as satisfactory, 3) the role of authority as a bar to the re-consideration of a problem. As regards the first of these, it should be emphasized that the process of tackling one problem by analogizing from another has frequently yielded valuable results, as in the case of air-pressure. On the other hand, it may lead to the adoption of a totally false hypothesis, as when the idea of the atom as an infinitely small piece of solid matter was obtained by analogizing from the world of visible appearances. This erroneous viewpoint blocked progress in this field for many decades. Similarly, the comparison of the movement of light to a wave—an analogy which had actually provided a satisfactory

explanation of the observed phenomenon during most of the nineteenth century—tended subsequently to interfere with the development of the equally valid concept of light as a stream of particles. This example also illustrates the second factor enumerated above. As far as the third factor is concerned, the history of science shows many instances in which the force of authority has operated in such a manner as to build up an exceedingly powerful resistance to further investigation; in some cases centuries elapsed before this resistance was eventually broken down, as happened in cosmology, for example.

[8] Thus in addition to the chances of going astray outlined, the scientific investigator shares with the ordinary citizen the possibilities of falling into errors of reasoning in the ways we have just indicated, and many others as well. The more he knows of this important subject, therefore, the better equipped he will be to attain success in his work; and the straighter he thinks, the more successfully he will be able to perform his function as a citizen.

分析：对这篇文章首先采用 skimming 或者 previewing 的方法获取了段落大意。每段的中心思想分别如下：

Paragraph 1. Reasons for reasoning incorrectly

Paragraph 2. Different understanding of words

Paragraph 3. Confusion of a word or a name with a fact

Paragraph 4. Mistakes in logic

Paragraph 5. An unreasonable inference

Paragraph 6. Confusion of cause and effect

Paragraph 7. Other factors of reasoning incorrectly

Paragraph 8. Conclusion

这篇文章的提纲十分清楚，具体如下：

I. Introduction

  II. Main reasons for reasoning incorrectly

    A. Misuse of words

      1. Different understanding of words

      2. Confusion of a word or a name with a fact

    B. Mistakes in logic

Unit Three   History of Science in Early Cultures

    1. Unreasonable inference

    2. Confusion of cause and effect

III. Sub-reasons for reasoning incorrectly

    A. Faulty analogizing

    B. Effect on further research

    C. The role of authority as a bar to the re-consideration

IV. Conclusion

以上例子充分说明，好的读者应该有相当的灵活性，根据不同的阅读目的，该快则快，该慢则慢，该精则精，该泛则泛，不能只用一种千篇一律的方法。要想真正掌握科技文献的阅读方法，还需要我们在今后的阅读中不断实践。

Quiz

I. Skim or preview the following passage, and write down the main idea of each paragraph.

Television: The Modern Wonder of Electronics

[1] Television, or TV, the modern wonder of electronics, brings the world into your own home in sight and sound. The name television comes from the Greek word "tele", meaning "I see". In Great Britain, the popular word for television is "telly".

[2] Television works in much the same way as radio. In radio, sound is changed into electromagnetic (invisible light) waves which are sent through the air. In TV, both sound and light are changed into electromagnetic waves. Experiments leading to modern television took place more than a hundred years ago. By the 1920s, inventors and researchers had turned the early theories into working models. Yet it took another thirty years for TV to become an industry.

[3] As an industry, TV provides jobs for hundreds of thousands who make TV sets and broadcasting equipment. It also provides work for actors, technicians, and others who put on programs. As an art, television brings the theater and other cultural events into the home. Its influence on the life of average Americans is incalculable; it can influence their thoughts, their likes and dislikes, their speech, and even their dress. It can also add to their store of knowledge. Through advertising, television helps businesses and manufacturers sell their products to millions of persons. Television has brought political campaigns closer to the voters than in former days. Educational TV stations offer teaching in various subjects ranging from home nursing to art appreciation. Many large schools and universities have "closed-circuit" television equipment that will telecast lectures and demonstrations to hundreds of students in different classrooms; and the lecture can be put on video tape to be kept for later use. Some hospitals use TV to allow medical students to get closed-up views of operations.

[4] In 1946, after World War II, TV began to burst upon the American scene with a speed unforeseen even by the most optimistic leaders of the industry. The novelty of seeing TV pictures in the home caught the public's fancy and began a revolution in the world of entertainment. By 1950, television had grown into a major part of show business. Many film and stage stars began to perform on TV as television audiences increased. Stations that once telecast for only a few hours a day sometimes telecast around the clock in the 1960s.

[5] With the development of programming also came the introduction of television in full color. By the middle 1960s, the national networks were broadcasting most of their programs in color. The obvious appeal of television, whether in color or black-and-white, can be documented by the increasing number of TV sets in homes around the country. By

the mid-1960s, 90 percent of the house-holds in the United States had at least one TV set, and 12 percent had two or more sets. TV had become a part of the daily life of the adults and children of America.

[6] The programs that people watch are not only local and national ones. Since the launching of the first communications satellite, more and more programs are televised "live" from all over the world. Television viewers in San Francisco were able to watch the 1964 Olympic Games in Tokyo by means of a communications satellite named Syncom. The Olympic Games in Mexico City and in Munich, Germany, were also telecast live, as were parts of the historic visit of President Nixon to the People's Republic of China. And live televisions now come from outer space; in 1969, the first astronauts to land on the moon televised their historic "moon talk" to viewers on the earth. Since then, astronauts have regularly sent telecasts to the earth.

[7] It look as if the use of television—in education, entertainment, and communication—appear to be endless. Certainly it is one of the major modern wonders of electronics in our changing world.

II. Scan the passage, and answer the following questions.

(1) By what date were the first TV working models constructed?

(2) By the mid-1960 what proportion of American homes had a TV set?

III. Read the passage carefully, and write down the outlines.

Part B Reading

Science in the Greek World

The history of science in early cultures refers to the study of protoscience in ancient history, prior to the development of science in the Middle Ages. In prehistoric times, advice and knowledge passed from generation to generation in an oral tradition. The development of writing enabled knowledge to be stored and communicated across generations with much greater fidelity.

Combined with the development of agriculture, which allowed for a surplus of food, it became possible for early civilizations to develop and more time to be devoted to tasks other than survival, such as the search for knowledge for knowledge's sake.

In Classical Antiquity, the inquiry into the workings of the universe took place both in investigations aimed at such practical goals as establishing a reliable calendar or determining how to cure a variety of illnesses and in those abstract investigations known as natural philosophy. The ancient people who are considered the first scientists may have thought of themselves as natural philosophers, as practitioners of a skilled profession (for example, physicians), or as followers of a religious tradition (for example, temple healers).

The earliest Greek philosophers, known as the pre-Socratics, provided competing answers to the question found in the myths of their neighbors: "How did the ordered cosmos in which we live come to be?" The pre-Socratic philosopher Thales[1] (640–546 B.C.), dubbed the "father of science", was the first to postulate non-supernatural explanations for natural phenomena: "All things are water". Thales' student Pythagoras[2] of Samos founded the Pythagorean School, which investigated mathematics for its own sake, and

Unit Three History of Science in Early Cultures

was the first to postulate that the Earth is spherical in shape. Leucippus[3] (5th century B.C.) introduced atomism, the theory that all matter is made of indivisible, imperishable units called atoms. This was greatly expanded by his pupil Democritus[4].

Subsequently, Plato[5] and Aristotle[6] produced the first systematic discussions of natural philosophy, which did much to shape later investigations of nature. Their development of deductive reasoning was of particular importance and usefulness to later scientific inquiry. Plato founded the Platonic Academy in 387 B.C., whose motto was "Let none unversed in geometry enter here", and turned out many notable philosophers. Plato's student Aristotle introduced empiricism and the notion that universal truths can be arrived at via observation and induction, thereby laying the foundations of the scientific method. Aristotle also produced many biological writings that were empirical in nature, focusing on biological causation and the diversity of life. He made countless observations of nature, especially the habits and attributes of plants and animals in the world around him, classified more than 540 animal species, and dissected at least 50. Aristotle's writings profoundly influenced subsequent Islamic and European scholarship, though they were eventually superseded in the Scientific Revolution.

The important legacy of this period included substantial advances in factual knowledge, especially in anatomy, zoology, botany, mineralogy, geography, mathematics and astronomy; an awareness of the importance of certain scientific problems, especially those related to the problem of change and its causes; and a recognition of the methodological importance of applying mathematics to natural phenomena and of undertaking empirical research. In the Hellenistic age scholars frequently employed the principles developed in earlier Greek thought: the application of mathematics and deliberate empirical research, in their scientific investigations. Thus, clear unbroken lines of influence lead from ancient Greek and Hellenistic philosophers, to medieval Muslim philosophers and scientists, to the European Renaissance and Enlightenment, to the secular sciences of the modern day. Neither reason nor inquiry began with the Ancient Greeks, but the Socratic Method did, along with the idea of Forms, great advances in geometry, logic, and the natural sciences.

The astronomer Aristarchus of Samos was the first known person to propose a heliocentric model of the solar system, while the geographer Eratosthenes [7] accurately calculated the circumference of the Earth. Hipparchus [8] (ca. 190–ca. 120 B.C.) produced the first systematic star catalog. The level of achievement in Hellenistic astronomy and engineering is impressively shown by the Antikythera mechanism [9] (150–100 B.C.), an analog computer for calculating the position of planets. Technological artifacts of similar complexity did not reappear until the 14th century, when mechanical astronomical clocks appeared in Europe.

In medicine, Hippocrates [10] (ca. 460–ca. 370 B.C.) and his followers were the first to describe many diseases and medical conditions and developed the Hippocratic Oath for physicians, still relevant and in use today. Herophilos [11] (335–280 B.C.) was the first to base his conclusions on dissection of the human body and to describe the nervous system. Galen [12] (129–ca. 200) performed many audacious operations—including brain and eye surgeries—that were not tried again for almost two millennia.

The mathematician Euclid [13] laid down the foundations of mathematical rigor and introduced the concepts of definition, axiom, theorem and proof still in use today in his Elements, considered the most influential textbook ever written. Archimedes [14], considered one of the greatest mathematicians of all time, is credited with using the method of exhaustion to calculate the area under the arc of a parabola with the summation of an infinite series, and gave a remarkably accurate approximation of Pi. He is also known in physics for laying the foundations of hydrostatics, statics, and the explanation of the principle of the lever.

Theophrastus [15] wrote some of the earliest descriptions of plants and animals, establishing the first taxonomy and looking at minerals in terms of their properties such as hardness. Pliny the Elder [16] produced what is one of the largest encyclopedias of the natural world in 77 A.D., and must be regarded as the rightful successor to Theophrastus. For example, he accurately describes the octahedral shape of the diamond, and proceeds to mention that diamond dust is used by engravers to cut and polish other gems owing to its great hardness. His recognition of the importance of crystal shape is a precursor to modern crystallography, while mention of numerous other minerals presages mineralogy. He also recognizes that other minerals have characteristic crystal shapes, but in one example, confuses the crystal habit with the work of lapidaries. He was also the first to recognize that amber was a fossilized resin from pine trees because he had seen samples with trapped insects within them.

New Words and Expressions

proto science 原型科学
fidelity 精确
the pre-Socratics 苏格拉底前的古希腊哲学家
cosmos 宇宙
dub 授予；给……起绰号
postulate 假定；假设
Samos 萨摩斯岛
atomism 原子论
indivisible 不可分割的
imperishable 不灭的，不朽的
Platonic 柏拉图式的，柏拉图学派的
motto 箴言，座右铭
empiricism 经验论，实证论
induction 归纳，归纳法
causation 因果关系
attribute 属性，特征
dissect 解剖（动、植物等）
supersede 代替，取代
legacy 遗产
methodological 方法（论）的；教学法的
secular 现世的，世俗的，非宗教的
heliocentric 日心的
circumference 圆周；圆周线
catalog 目录，条目
audacious 大胆的，有冒险精神的
millennia（millennium 的复数）一千年
axiom 公理，原理
theorem 定理
arc 弧
parabola 抛物线
summation 和；求和（法）

approximation 近似值；近似法	zoology 动物学
hydrostatics 流体静力学	botany 植物学
taxonomy 分类法	mineralogy 矿物学
successor 后继者，接班人，继承人	astronomy 天文学
octahedral 有八面的；八面体的	crystallography 结晶学
engraver 雕刻师	lapidary 宝石工艺匠
gem 宝石，美玉	amber 琥珀
anatomy 解剖学	resin 树脂

Notes

1. Thales 泰勒斯，古希腊时期的思想家、科学家、哲学家，希腊最早的哲学学派（米利都学派，也称爱奥尼亚学派）的创始人。

2. Pythagoras（前572—前497）毕达哥拉斯，古希腊数学家、哲学家。

3. Leucippus（约前500—约前440）留基伯，古希腊唯物主义哲学家和原子论的奠基人之一。

4. Democritus（约前460—前370）德谟克利特，古希腊哲学家。

5. Plato（约前427—前347）柏拉图，古希腊哲学家。

6. Aristotle（前384—前322）亚里士多德，世界古代史上最伟大的哲学家、科学家和教育家之一。

7. Eratosthenes（公元前3世纪）埃拉托色尼，古希腊天文学家、数学家和地理学家。

8. Hipparchus（约前195—前125）希帕克（或译为喜帕恰斯），古希腊天文学家，被称为"方位天文学之父"。

9. Antikythera mechanism 安提基特拉机械是为了计算天体在天空中的位置而设计的古代青铜机器，属于模拟计算机。该机器是于1901年在希腊安提基特拉岛附近的安提基特拉沉船里发现的，发现之后有数十年间都不明白其内部复杂构造。该机械的制造年代约在西元前150到西元前100年之间。直到14世纪欧洲天文钟才重新体现如此复杂的工艺技术。

Unit Three   History of Science in Early Cultures

10. Hippocrates 希波克拉底，古希腊医师，被誉为西方"医学之父"。Hippocratic Oath 为希波克拉底誓词，俗称医师誓词，是西方医生传统上行医前的誓言。在希波克拉底所立的这份誓词中，列出了一些特定的伦理上的规范。不过希波克拉底本人可能不是这份誓词的原作者。

11. Herophilos（前335—前280）赫洛菲洛斯，第一位用有条理的技巧进行人体解剖的学者。

12. Galen 盖伦，希腊解剖学家、内科医生和作家，其著作对中世纪医学有决定性影响。

13. Euclid（前330—前275）欧几里得，古希腊数学家。

14. Archimedes（前287—前212）阿基米德，古希腊哲学家、数学家和物理学家。

15. Theophrastus 泰奥弗拉斯托斯，古希腊哲学家和自然科学家。

16. Pliny the Elder（23—79）老普林尼（或大普林尼），古罗马作家、博物学者、军人、政治家。以《自然史》(*Naturalis Historia*，又译《博物志》)一书留名后世。其养子为小普林尼（Pliny the Younger）。

Exercises

I. Find out the English equivalents to the following Chinese terms from the passage.

1. 古典时代
2. 自然哲学
3. 自然现象
4. 演绎推理
5. 科学革命
6. 希腊化时代
7. 欧洲文艺复兴和启蒙运动
8. 苏格拉底问答法
9. 希波克拉底誓言
10. 数学严密性
11. 穷举法
12. 晶体惯性

II. Choose the best answer according to the information in the passage.

1. The first scientists thought of themselves as the following except _____.
   A) natural philosophers

B) practitioners of a skilled profession

C) followers of a religious tradition

D) the pre-Socratics

2. Who was the first person to postulate that the Earth is spherical in shape?

   A) Thales   B) Pythagoras

   C) Leucippus   D) Democritus

3. Which statement is NOT true about Aristotle?

   A) He was Plato's student.

   B) He believed that universal truths can be arrived at via observation and induction.

   C) He produced many biological writings that were empirical in nature.

   D) His writings profoundly influenced subsequent Scientific Revolution.

4. The great achievement in Hellenistic astronomy and engineering is _____.

   A) a heliocentric model of the solar system proposed by Aristarchus

   B) the accurate calculation of circumference of the Earth calculated by Eratosthenes

   C) an analog computer used for calculating the position of planets

   D) the first systematic star catalog produced by Hipparchus

5. Which of the following is NOT true about Archimedes?

   A) Archimedes laid down the foundations of mathematical rigor.

   B) Archimedes used the method of exhaustion to calculate the area under the arc of a parabola.

   C) Archimedes gave a remarkably accurate approximation of Pi.

   D) Archimedes explained the principle of the lever.

Part C Extended Reading

Plato and Aristotle

Plato (pointing up to heavenly things) and Aristotle (gesturing down to Earth).
From Raphael, *The School of Athens* (1509)

Like the Pythagoreans, Plato (c. 427–c. 347 B.C.) found the ordering principle of the universe in mathematics, specifically in geometry. A later account has it that Plato had inscribed at the entrance to his school, the Academy, "Let no man ignorant of geometry enter." The story is a myth, but like all myths it has a grain of truth, for in his writings Plato repeatedly tells us of the importance of geometry.

Plato is known more for his contributions to the philosophical basis of scientific method than to particular scientific concepts. He maintained that all things in the material world are imperfect reflections of eternal unchanging ideas, just as all mathematical diagrams are reflections of eternal unchanging mathematical truths. Since Plato believed that material things had an inferior kind of reality, he considered that we don't achieve demonstrative knowledge—that kind of knowledge we call science—by looking at the imperfect material world. Truth is to be found through rational demonstrations, analogous to the demonstrations of geometry.

Applying this concept, Plato recommended that astronomy be studied in terms of geometrical models and proposed that the elements were particles constructed on a geometrical basis.

Aristotle (384–322 B.C.) disagreed with his teacher, Plato, in several important respects. While Aristotle agreed with Plato that truth must be eternal and unchanging, he maintained that we come to know the truth through the external world which we perceive with our senses. For Aristotle, directly observable things are real; ideas (or as he called them, forms) only exist as they express themselves in matter or in the mind of an observer or artisan.

This theory of reality led to a radically different approach to science:

- First, Aristotle emphasized observation of the material entities which embody the forms.
- Second, he played down the importance of mathematics.
- Third, he emphasized the process of change where Plato had emphasized eternal unchanging ideas.
- Fourth, he reduced the importance of Plato's ideas to one of four causal factors.

As this last point suggests, Aristotle's concept of causes was less limited than ours. Among causes he included:

- The matter of which a thing was made (the material cause).
- The form into which it was made (the formal cause; something similar to Plato's ideas).
- The agent who made the thing (the moving or efficient cause).
- The purpose for which the thing was made (the final cause).

Aristotle's emphasis upon causes fundamentally shaped the later development of science by insisting that scientific knowledge, what the Greeks called episteme and the Romans scientia, is knowledge of necessary causes. He and his followers would not accept mere description or prediction as science. In view of this disagreement with Plato, Aristotle established his own school, the Lyceum, which further developed and transmitted his approach to the investigation of nature.

Most characteristic of Aristotle's causes is his final cause, the purpose for which a thing is made. He came to this insight through his biological researches, in which he noted that the organs of animals serve a particular function.

Thus Aristotle was one of the most prolific natural philosophers of Antiquity. He made countless observations of nature, especially of the structure and habits of plants and animals. He also made many observations about the large-scale workings of the universe, which led to his development of a comprehensive theory of physics. For example, he developed a version of the classical theory of the elements (earth, water, fire, air, and aether). In his theory, the light elements (fire and air) have a natural tendency to move away from the center of the universe while the heavy elements (earth and water) have a natural tendency to move toward the center of the universe, thereby forming a spherical earth. Since the celestial bodies—that is, the planets and stars—were seen to move in circles, he concluded that they must be made of a fifth element, which he called Aether.

Aristotle could point to the falling stone, rising flames, or pouring water to illustrate his theory. His laws of motion emphasized the common observation that friction was an omnipresent phenomenon—that any body in motion would, unless acted upon, come to rest. He also proposed that heavier objects fall faster, and that voids were impossible.

New Words and Expressions

Pythagorean 信奉毕氏学说者
inscribe 刻（写）；雕
Academy 学园（柏拉图讲哲学的地方）
a grain of 些微，一点儿
demonstrative 论证的
analogous 类似的；可比拟的
entity 存在，实体
play down 降低，贬低，减弱
Lyceum 学园（亚里士多德讲学的地方）
prolific 多产的；富于创造力的
aether (ether) [物] 以太，能媒
celestial body 天体
omnipresent 无所不在的
void 空位；空间；真空

Exercises

Answer the following questions by using the sentences in the passage or in your own words.

1. What did Plato emphasize?

2. What is Plato known more for?

3. Did Aristotle agree with Plato?

4. What was the result of Aristotle's theory of reality?

5. What's the influence of Aristotle's emphasis on causes?

Unit Four

History of Science and Technology in Ancient China

Part A　Lecture

确定中心思想

中心思想也称作主题思想，是作者在文章（或段落）中要表达的核心内容，因而也是读者在阅读中所应获取的最重要的信息。能否在阅读文章（或段落）时抓住中心思想，体现了读者总结、概括和归纳事物的能力。把握中心思想也有助于对文中具体内容的理解，因为文中的词、句都围绕并服务于阐述中心思想。因此，确定中心思想是一项最重要的阅读技巧。本单元讲座将阐述确定中心思想的技巧。

1　确定中心思想的技巧

每篇文章（或段落）都有一个主题，这是作者向读者说明的一个道理或阐述的一个观点。因此，把握文章（或段落）的主题是最根本的。一篇文章（或一个段落）可以有很多次要思想（minor ideas）和辅助细节（supporting details），但中心论点或中心思想一般只有一个。而这个中心思想是作者在该文章（或段落）中所要传递的最重要的思想，这一思想贯穿文章始末，渗透于每个段落、每个句子。中心思想起主导作用，作为论据的次要思想和辅助细节是主题的从属，它们之间是"主干"与"分枝"的关系。阅读过程中，如果不厘清这种关系，就会导致判断错误。试分析以下短文，找出中心思想。

All students of geography should be able to interpret a weather map accurately. Weather maps contain an enormous amount of information about weather condition existing at a time of observation over a large geographical area. They reveal in a few minutes what otherwise would take hours to describe. The United States Weather Bureau issues information

Unit Four History of Science and Technology in Ancient China

about approaching storms, floods, frosts, droughts, and all climatic conditions in general. Twice a month it issues a 30-day "outlook" which is a rough guide to weather conditions likely to occur over broad areas of the United States. These 30-day outlooks are based upon an analysis of the upper air levels which often set the stage for the development of air masses, frosts and storms.

What is the main idea of this passage?

A) Weather maps are very important to students of geography.
B) Weather maps can have a 30-day "outlook".
C) Students of geography should be able to make weather maps accurately.
D) Those who major in geography should possess the ability to have a good understanding of a weather map.

分析：纵观例 1 全文，我们发现，短文谈论的主题是气象图。就短文的结构而言，第 1 句为主题句，第 2 句开始至最后一句都是围绕着第 1 句这个中心来加以解释的。作者从以下几方面进行解释：一是气象图信息包容量大；二是覆盖面大；三是速度快；四是可作为期三十天的气象展望，还可有效地预报灾难性天气。而作为一个学地理的学生，看懂这样的气象图是必不可少的。所以原文第 1 句和下文是一种主从关系。了解了这一点，就不难获取本段的中心思想。上述选项中 D 项与原文的主题句相吻合。

从以上分析可以看出，中心思想就是表达统帅全篇（段）思想的、能涵盖全篇（段）各句所包含意义的句子。通过文章（或段落）的结构、主题和主题句是获取中心思想的一种形式。获取文章（或段落）中心思想的另一种形式是为文章（或段落）选择合适的标题，因为标题常常也体现它所围绕的主题或讨论的中心。此外，判断作者或文章的意图也可有效归纳出文章（或段落）"暗藏"的中心思想，因为作者或文章的意图与主题、主题句和中心思想是一致的。

以下我们将讨论如何利用篇章结构、主题、主题句、标题和作者或文章意图确定中心思想。

1.1 利用篇章结构确定中心思想

作者按一定规律组句成章。篇章结构体现作者的谋篇布局和思想脉络。因此,掌握篇章结构的基本规律,对于确定篇章的中心思想,了解作者阐述中心思想的构思,从篇章整体上促进对文章的理解是很有帮助的。

我们知道,具有完整结构的说明文通常包括引言(导言)、正文和结束语三个部分。引言是文章的起始部分,用于点明主题,并规定围绕该主题展开文章的方式,以开启下文。正文是文章的主体部分。正文依照引言规定,对主题进行阐述。正文部分往往包含一个以上的段落。每个完整的段落都围绕一个主题进行阐述。段落中通常有表达中心思想的主题句和围绕主题句展开的辅助句。有的段落还有总结句对主题句进行重述,以示强调。段落所阐述的内容往往是按一定的逻辑关系排列的。不同的逻辑排列形成不同的段落类型。结束语是文章的收尾部分。不是每篇文章都有结束语,但好的结束语能起到照应开头和正文,使文章首尾一致的作用。篇章结构见图 4-1。

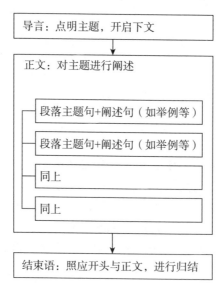

图 4-1 篇章结构框图

下面我们试分析一篇规范运用篇章结构基本模式的短文。

Unit Four History of Science and Technology in Ancient China

(1) Vilma likes traveling by bus better than by airplane for four reasons. (2) First, it costs less. (3) She can ride by bus from New York to Chicago for ＄20. (4) Yet, the air fare between the two cities is ＄78. (5) Second, traveling by bus gives her a closer look at the cities and countryside than she could get from a plane. (6) For example, she enjoys driving through the big cities of Philadelphia, Pittsburgh, and Fort Wayne. (7) Third, Vilma finds that the passengers on the bus are often closer to her own age. (8) So, they are easy to talk with on the shared travel adventure. (9) In contrast, many airline passengers are businessmen who keep busy with their work while flying. (10) Fourth, riding a bus allows Vilma to explore any stop along the way. (11) For instance, once she got off the bus at a small town in Indiana and spent the night with a girl friend. (12) Then, the next day she took another bus to Chicago. (13) For these reasons, Vilma usually prefers buses instead of planes unless she has to get somewhere in a hurry.

分析：这篇短文思路清晰，结构完整。短文分成引言、正文和结束语三部分。第 1 句为引言，点明了全篇的主题，并通过介词短语 for four reasons 规定了正文展开的方式是列举四条理由。第 2 至 12 句是正文，对主题进行阐述：句 2 给出理由一，句 3 与句 4 通过对比支撑句 2；句 5 给出理由二，句 6 为例证；句 7 给出理由三，句 8 与句 9 使用对比解释理由三；句 10 提出理由四，由句 11 与句 12 的实例支撑。最后一句为结束语，通过介词短语 for these reasons 照应开头与正文，重述主题，与开头呼应。短文的篇章结构见图 4-2。

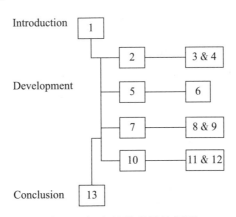

图 4-2 短文的篇章结构框图

从上例可以看出，短文的篇章结构安排直接反映了作者的思路脉络。因而利用篇章结构规律确定文章的中心思想是一种较可靠的方法。通常可以采用以下步骤：1）阅读文章的引言部分。找到点明主题的句子即找到了中心思想。2）阅读正文中各段的主题句，这些主题句所围绕的中心就是文章的中心思想。3）阅读结尾段（或句），找到与文章开头呼应的、对主题进行重述的句子，可以验证已找出的中心思想是否正确。

1·2 利用主题确定中心思想

文章的主题是文章的纲，在阅读中如能抓住这个纲，就能纲举目张。确定文章主题的能力是我们应掌握的技能。

文章的主题（topic 或 subject）指文章所讨论的主要论题。一篇（段）文章，可能会涉及好几个话题，但其中必定有一个是主要的、中心的话题，称为主题（main subject 或 main topic），其余都是次要话题，为主要话题铺垫服务的。试分析下面短文，并找出主题。

Magnesium（镁）is another mineral we now obtain by collecting huge volumes of ocean water and treating it with chemicals, although originally it was derived only from brines（海水）or from the treatment of such magnesium-containing rocks as dolomite（石灰石）of which whole mountain ranges are composed. In a cubic mile of seawater there are about four million tons of magnesium. Since the direct extraction method was developed about 1941, production has increased enormously. It was magnesium from the sea that made possible the wartime growth of the aviation（航空）industry. Every airplane made in the United States (and in most other countries as well) contains about half a ton of magnesium metal. And it has innumerable uses in other industries where a lightweight metal is desired, besides its longstanding utility as an insulating material, and its use in printing inks, medicines, and toothpastes.

What is the main topic of this passage?

    A) Uses of seawater

Unit Four History of Science and Technology in Ancient China

   B) Treatment of seawater

   C) Chemical properties of magnesium

   D) Derivation and uses of magnesium

   E) Aviation industry

分析：本短文论及两方面内容：一是镁的提炼方法、发展等，二是镁的用途，即造飞机，促进航空工业发展及在其他工业方面也有许多用途。我们从问题中列举的选项可知，各项都是文中谈到的话题。因此，我们要判断哪一个话题是可以包含其余话题的总题，或者说其余的话题可以归结到哪一个总题上去。A 项和 B 项是关于海水用途及海水处理，实际同属一个问题，两项内容合并构成文章第 1 句的部分内容：海水被收集处理提取镁。可见，A、B 都不能作为概括全篇的主题。C 项谈及镁的化学性质，而文中并未提及镁的化学性质，只提到镁的重量（lightweight），属物理性质，所以 C 也不是主题。D 项谈及镁的提炼和用途，可以概括短文谈及的两个方面内容。而 E 项只涉及镁的用途的一个局部性问题，从属于 D 项。综上所述，D 项所谈论的话题最具包容性，是短文的主题。

最具包容性的话题往往也是那些最根本、最具有概括力的信息，常常就包含在主题句或结论句里。

　　值得注意的是，文章主题与中心思想（或主题思想）是有区别的。主题是用词组高度概括的论题范围；中心思想是用句子反映文章的主要思想，包含作者的观点意见或结论性看法。

　　The energy crisis the world is now experiencing has prompted scientists to investigate new sources of energy. Solar power is receiving wide attention as a possible source of clean, inexpensive and abundant energy. The conversion of the sun's rays to electricity does not pollute the environment and costs only as much as the installation of solar panels to absorb the sun's rays. Admittedly, this system would not work well in a cloudy area, but in countries like the United States where sunny days are numerous, solar power could help solve the energy crisis.

(1) The central thought of the paragraph deals with _____.

   A) a problem and its possible solution

B) the cause of the world's energy shortage

C) the effects of energy mismanagement

D) the financial benefits of solar heat

(2) Which sentence best expresses the main idea?

A) The energy crisis has encouraged researches into new sources of energy.

B) The use of solar energy is the only solution to the energy crisis.

C) Solar power is one of the cleanest sources of energy.

D) The changing climatic conditions around the world limit the possibilities of solar energy.

分析：本文谈及能源危机问题促使科学家探讨解决的方法。

题 1 的答案为 A，从短文可知，作者提出了一个问题（energy crisis），也提出了一个解决办法（solar energy）；表达主题是用短语形式。

题 2 的答案为 A，因为它概括了本文所叙述的主旨，指出了各个句子的发展方向；表达中心思想是用句子表述文章的主要内容。

1.3 利用主题句确定中心思想

正文的每个段落都围绕文章的主题，说明主题的某个方面。通常情况下，每个段落都有一个概括说明本段中心思想的关键句，称为主题句。段落中的其余句子称为辅助句，用来解释、支撑或扩展主题句所表达的中心思想。在阅读文章时，能否找出主题句并充分利用它来确定中心思想是很关键的。

一般说来，每篇文章（或每个段落）都有其主题句。主题句的位置可出现在段首、段末或段中，而以位于段首最为常见。

1.3.1 主题句位于段首

这是主题句最常见的位置，在有主题句的段落中占 60% 以上。作者开门见山点出段落的主旨，之后用事例、细节加以阐释。这样的段落在逻辑结构上符合演绎法从一般到个别的写作程序，可用一个倒置的三角形（▽）表示。其他句子通常由主到次按句子的重要性或与主题句的关系密切性进行排列。

Unit Four History of Science and Technology in Ancient China

Clouds can greatly affect the temperature of the earth's surface. When there are many clouds in the sky, all of the sun's rays cannot reach the earth. The cloudy day, then, will be cooler than the cloudless day. Clouds also prevent the earth from cooling off rapidly at night. For this reason, countries such as the British Isles, which are often covered by clouds, have a relatively constant temperature. The weather in these cloudy areas is neither very hot in summer nor very cold in winter. On the other hand, places such as deserts, which have few or no clouds, have very sharp variations in temperature—between night and day as well as between summer and winter.

分析：短文第 1 句是主题句，提出论点：云能显著影响地表温度。其余的句子都是辅助句。其中第 2 至 4 句从两个方面（白天云能遮阳、夜间云能阻止散热过快）给出论据，支撑主题句。句 5 至句 7 则通过具体实例的对比进一步加以证实。句 5、句 6 说明多云地区相对恒温：昼夜温差小，冬暖夏凉。句 7 对比无云的沙漠地区温差极大。所有的辅助句从不同的角度阐述了主题。

1.3.2 主题句位于段尾

当主题所代表的内容比较复杂且不太为读者所熟悉，或写作目的是劝导与说理时，为了使这个复杂的主题更易于读者理解；或为了使读者更易于接受作者的观点，有必要先把解释性的细节或接受这个观点的理由说个透彻，这时通常就把内容较复杂或代表作者观点的主题句置于段尾位置，这样的段落在逻辑结构上则是表达细节的句子在前、概述性句子居后，符合归纳法从个别到一般的写作程序，可用一个正三角形（△）表示。

Like the airplane, the bird has wings. Both use power to pull themselves through the air. The airplane uses a propeller to do this. Birds do it by beating the air back with their wings. Both airplanes and birds have streamlined bodies so that they can slide more easily through the air. They both use their tails as rudders to steer upward or downward. **Birds are really like airplanes in many ways.**

分析：这段文字对鸟和飞机的飞行方式做了详细的比较，然后得出结论：鸟在许多方面与飞机有着相似之处。主题句为最后一句话。

▶ 1.3.3 段首 + 段尾主题句

有时为了特别强调主题句的内容，作者把主题句同时置于段首和段尾，做到前后呼应，重复点题。尾句重述首句所表述的思想，但又绝非简单的重复，更多的是尾句有所引申，并顺段落细节的铺叙另有侧重。这种段落在逻辑结构上从一般到个别，再从个别到一般，先演绎再归纳，可用一个正三角形上面叠一个倒三角形（⧖）表示。

 例 7

Good manners are important in all countries, but ways of expressing good manners are different from country to country. Americans eat with knives and forks; Japanese eat with chopsticks. Americans say "Hi" when they meet; Japanese bow. Many American men open doors for women; Japanese men do not. On the surface, it appears that good manners in America are not good manners in Japan, and in a way this is true. But in any country, the only manners that are important are those involving one person's behavior toward another person. In all countries it is good manners to behave considerately toward others and bad manners not to. **It is only the way of behaving politely that differs from country to country.**

分析：段首点明主题"良好举止因国而异"，接着举例说明各国的差异，段尾进行归结，用强调句式重述这一主题。

▶ 1.3.4 主题句位于段落中间

有的段落的主题句既非第一句，也非最后一句，而是出现在一段的中间。在这样的段落中，作者先引题后演绎，由细节到主题，再由主题回到细节，在逻辑结构上包括三个层次：引题—主题—阐述，可用菱形（◇）表示。

Unit Four History of Science and Technology in Ancient China

Despite the fact that cars from Germany and Japan are flooding the American market, **Ford, General Motors and Chrysler are hiring more workers than ever before. The flood of cheaper foreign cars has not cost American auto workers their jobs as some experts predicted.** Ford operates as far as Asia, and General Motors is considered Australia's biggest employer. Yet GM has its huge American work force and hires hundreds of people every day to meet the needs of an insatiable society.

分析：本段的主题句为第 2 句。句 1、句 3、句 4 都是具体阐述中心思想的辅助句。

▶ 1.3.5 主题句隐含在段落的内容中

有的段落没有明确的主题句，但文中所有的事实和细节形成一个整体，整个段落由一个隐含其中的思想所统一。读者必须综合文中的例证，进行正确推理才能归纳出统领全段的主导思想。这种隐含主题的推理可从两方面进行：一是确定全段论述的主要对象（何人、何物、何事、什么观点等）；二是确定论述该主要对象的哪个主要方面（行为、状态、作用、特征、功能、问题、优劣等）。这种段落的逻辑结构，可用一个矩形（□）表示。

Low temperatures slow down the rate of chemical and physical processes in the body. This fact can be used to advantage by a surgeon who intends to perform a heart operation which will take a long time. If he cools down the body, the brain will require less oxygen from the blood. This will make the operation much safer, since during an open heart operation the brain must of necessity be deprived of its normal supply of fresh blood.

分析：此段没有陈述主题思想的句子，且文中四句的重要性呈平行状态。句子的重心未显示出任何的方向性。四句形成一个整体，构成作者要表达的中心思想。主题句可以归纳为：Cooling can help surgeons in heart operation.

1.4 利用文章标题确定中心思想

标题(title)是一篇文章的题目。文章的标题最具涵盖性,即常用鲜明醒目、概括性的词语来表明文章的主要意思,因此最能集中体现全篇(段)的中心思想。

在阅读过程中,如果有标题的话,适当参考文章(或段落)标题,会大大有助于确定文章的主题思想。且看下例。

Subjects to Avoid in Western Countries

There are certain things which it is considered bad manners in western countries to talk about in society. It is important to know these and avoid them. The subjects to be avoided are: bodily functions, or anything connected with more private parts of the body, details of birth, details of unpleasant illnesses; income of friends, or prices of their possessions: the age of the person one is talking with; personal questions or remarks, such as, "why don't you get married?" or "I should think you would like to have some children". Some of these are allowed in Chinese society, but **they are all taboo in formal western society.**

分析:本段的第 1 句是主题句,接着分述要避免的一些话题,段尾重述指出这些话题是禁忌。根据主题句和最后的重述性总结句,再参考标题,很容易确定短文的中心思想:It is important to know all taboo in formal western society。

在阅读过程中,如果没有标题,我们也可为文章(或段落)选择合适的标题,因为标题常常也体现它所围绕的主题或讨论的中心。因此,确立文章(或段落)标题实际上也反映了对中心思想的掌握。一般说来,确立标题有两个步骤:1)寻找主题句;2)选取与主题句中的关键词或主题词相近或相同的词或词语做标题。

Honeybees cannot live alone. Their body structure and instincts equip them for life in a colony or community, where they have a complex social organization and the various duties are divided among the individual

Unit Four History of Science and Technology in Ancient China

according to physical fitness and age. An individual worker bee cannot reproduce itself. While it may continue to live if forcibly isolated from its mates, it fails to care for itself adequately, and soon dies. Most insects have the ability to hibernate in winter, but the honeybee seems to have lost this. Since at low temperatures the bee will die, it must have the ability to make its own environment, so far as temperature is concerned. This makes a colony necessary to the bees in winter, so that they may collectively warm each other. Efficiency, if not necessity, demands that the work of the colony be divided, and such labor tends to enhance the need to maintain the colony. The physical structure of the honeybee is further suited for the defense of the entire colony rather than for its own defense. The bee's barbed sting is used only once and is made more effective by the fact that it is left behind in the victim. With the loss of the sting, however, the bee dies. This kind of defensive weapon is not of service to the individual, but to the community.

Which of the following is the most appropriate title for the passage?

A) The Communal Life of Bees B) The Structure of the Bee

C) The Organization of Insect Colonies D) The Life of Social Insects

分析：短文第 2 句是主题句。第 1、2 句从一正一反两个方面说明蜜蜂适合集群生活（Life in colony or community），即 communal life，全文的主题是讲蜜蜂的集群生活。主题与标题在本文被混同看待，所以答案选 A。

其实，主题和标题是不同的概念。主题是文章的中心论题，是一种客观表述形式；而标题是有倾向性地表现主题的一种主观表达形式，标题除有与主题相同的简洁、鲜明、概括等特点外，还具有主题所不具有的其他特点，如艺术性、暗示性、招揽性和可读性等。

There is a simple economic principle used to determine prices. It is called the law of supply and demand. Supply means the amount of or access to, a certain goods. Demand represents number of people who want those goods. If the demand for those goods is much greater than the supply, then the price rises, of course, manufacturers prefer to sell more goods at increased prices.

Which is the best title?

A) Economic Principles

B) Law of Supply and Demand

C) More Goods, Lower prices

D) Fewer Goods, Higher prices

分析：这段文章的中心论题是供需规律。第 2 句是主题句，后面各句解释供需规律及该规律引起的价格变化。这里要求选标题，实际就是确立主题，所以选 B。B 项能准确表现主题，此处标题与主题具有一致性。

在确立标题时，常常可能遇到没有主题句的篇章。这时，我们就要概括出主题，然后用与主题相关的概括词或词组做标题。

In modern society, it is often difficult for people to meet one another. Men and women of similar interests or needs may be separated from one another by their jobs or by great distances. As a result, they must advertise. Open any newspaper and you will find personal ads of all sorts. Lonely men and women advertise for husbands, wives, lovers, traveling companions, and partners both in business and in personal relationships. There are also ads for services that provide computer-chosen dates or escorts（伴随者）. Are you looking for someone new in your life? Try looking the personal ads in your newspaper. It is filled with opportunity.

Choose the best title for the above passage.

A) Difficulty for People to Meet

B) Newspaper

C) Ads

D) Meeting People through the Newspaper

分析：这段文章没有主题句。全文各句的焦点是报纸，通过报纸（广告）提供机会，解决人们相见的难题（difficulty for people to meet）。所以，中心思想是：The newspaper solves difficulty of people's meeting in modern society; 主题是：The difficulty of people's meeting and the newspaper; 标题是：Meeting People through the Newspaper。

Unit Four History of Science and Technology in Ancient China

从以上例子可以看出，标题实际上有两层含义。1）在某些场合，标题与主题混用，确立标题实际就是确立主题；2）在某些场合，标题与主题有区别：主题是对题材的客观概括，标题是具有作者思想倾向的主观性的表达，有时具有艺术性，有一定招揽性。

1·5 通过判断作者意图或文章意图确定中心思想

作者写一篇文章，或是要陈述一个事实，或是要解释一个问题，或是要论证一个道理……这种"陈述""解释""论证"就是作者的意图。作者的意图通过文字语言手段体现出来，就是文章意图。考核对作者（或文章）意图的理解，与考核对文章主题、主题思想的理解以及对文章标题的确立如出一辙，即考核对文章主题主旨的理解，只是考核的角度和方式稍有不同罢了。

Convective flow（对流）should be familiar to anyone who has noted the boiling of a heated liquid. The most elementary type of convection can be explained by the fact that heat rises. In the simplest cases, convective flow begins when a fluid is heated from below. As the bottom layer of the fluid is heated, it expands and thus becomes less dense than the layers above. The warmer and lighter bottom layer then tends to rise and the cooler layer tends to sink in a continuous cycle. The same mechanism of convective flow is responsible for the great ocean currents and for the global circulation of the atmosphere. In an ocean, the water is warmed by the sun to a depth of perhaps thirty meters, and evaporation near the water's surface is responsible for the cooling effect.

The main purpose of this passage is to _____.

  A) explain the basic principle of convection

  B) describe regular changes in the Earth's atmosphere

  C) explain the boiling temperatures of liquids

  D) state the principles of ocean current

分析：本文考核对文章/作者意图的理解。从短文中知道，第 2 句是主题句，说明文章的目的是要用热上升的事实来解释（explain）对流（原理）。本文的主题是 mechanism of convective flow；主题思想是 The mechanism of convective flow causes changes of atmosphere and ocean current. 文章的目的与主题句、主题、主题思想是一致的，故答案选 A。

我们知道，作者或文章意图是利用写作手段对题材进行分析、论证、加工处理而实现的。要判断作者或文章的意图，必须客观地分析文章的思路、作者加工处理素材的步骤和走向。一般说来，可以从以下两个方面判断作者或文章的意图：

▶ 1.5.1 抓住主题句和具有概括性的句子

有些文章的总括性、纲领性句子帮助读者了解作者思路，使读者能较容易地领会文章主旨，判断文章和作者的意图。如：

Although commonly thought to be one of man's feared deep-sea enemy, the octopus（章鱼）is generally a harmless animal that rarely bothers man. In fact, most of these powerful, eight-armed creatures are afraid of people. There have been some cases in which octopuses have attacked divers. But even these scattered attacks have not been very serious. In the mid nineteenth century, Hugo is said to have started the idea that the octopus is a vicious monster of the deep. In this "Toilers of the Sea" he described how this devil fish eats a human being. The tale became so popular that other novelists, and later the movies used episodes which describe man struggling in the arms of this frightening monster. And thus, the misconception of the octopus as a vicious, merciless killer was spread.

The purpose of this passage is to _____.

    A) describe the stories of the octopus

    B) explain the causes of the misconception of the octopus

    C) prove the harmlessness of the octopus

    D) correct the misconception of the octopus

分析：短文第 1 句是总括性句子，提出：1）octopus（章鱼）一般来说是

Unit Four History of Science and Technology in Ancient China

无害的动物（harmless animal），这是作者的观点；2）人们普遍认为章鱼是可怕的深海动物——这是作者退一步承认的事实。下面的文章就沿着第 1 句规定的思路写原因。1）为了说明章鱼无害，作者举了两条理由：A. 它几乎不害人，它怕人；B. 少数情况下攻击人，但不严重。2）为了说明人们普遍认为章鱼可怕，也有两方面原因：A. 作者 Hugo 的描述；B. 其他小说和电影的描写。所以根据以上的思路分析，我们很容易了解文章的观点和意图都在第 1 句里。文章的目的是消除人们对章鱼的误解，说明章鱼并非人们想象的那么可怕。故选答案 D。

▶ 1.5.2 采取综合归纳手段，找出主题思想

有些文章，全文（段）都在描写事物，或人物，或地点，或过程，没有主题句，但作者或文章的意图都隐含在字里行间。这时，我们必须纵观全局，找出隐含的主题思想，因为主题思想就是作者或文章要实现的意图。

 例16

Philip, a recent school graduate, has returned to his school to a former teacher. He appears extremely confused while he is talking about his serious dilemma at college. High school was a relatively successful experience for him. He achieved honor roll status（优等生）and received numerous athletic awards. In fact, he was given a full-tuition athletic scholarship to college he is now attending.

As his first semester at college progresses, Philip is finding it impossible to complete his school work and fulfill his obligations as an athlete at the same time. The demands of his professors are mentally exhausting, and he is worn out physically. It seems that the only way he can survive at college and maintain academic and athletic status is to cheat. Every other member of the team has cheated in one way or another. Crib（夹带）notes are available for most courses.

This bothers Philip. Where is he to draw the line?

The purpose of the passage is to _____.

A) describe Philip's obligations at college

B) show the teammate's influences on Philip

103

C) discuss Philip's trouble at college
D) advise students to practice crib if possible

分析：本文是一段情节性的文章，叙述刚上大学一年级的 Philip 返回中学拜访一位中学老师，在谈及大学里的窘境时显得很慌乱。短文追述了他在中学时风光情景——优等生，体育方面获过奖，并获得升大学的奖学金……可是，在大学第一学期他就发现不能完成学业和履行体育方面的义务，他感到教授们的要求很费脑筋，他已筋疲力尽，似乎只有作弊才能混下去。体育队里的队员们都以种种方式作弊，大多功课都搞夹带。这些事困扰着 Philip。他不知该怎么办？其实这就是他在大学里的窘境。纵观全文，短文讲述了 Philip 在中学里的 success 和大学里的 dilemma。从段末提出的问题看，重点是他面临的窘境，所以选 C。

2 综合练习

我们学习了以上确定中心思想的技巧，但是要想迅速而准确地识别文章的主题、找出文章的主题句、确立文章的标题、判断作者或文章的意图或抓住文章的主题思想，还必须通过大量的练习，才能学会识别文章中最根本、最具概括性的信息。在此过程中，通常可以采用略读的方法。

略读主要是用以捕获文章中心思想的阅读方法。略读时，从头至尾浏览全文，不要被个别生词或句子难住而停止下来，要特别注意读文章的首尾段和各段的首尾句，以及由 but、What's more、on the other hand 等逻辑词引出的句子；要看看全文谈论什么主题，作者从哪几个方面进行阐述的。为了更有效进行捕获中心思想的略读，我们可以带着以下问题有的放矢地阅读：1）文章（或段落）中大多数句子所共有的思想是什么？ 2）是什么把这些句子联系在一起？ 3）这些句子共同支持什么观点？ 4）这些句子共同阐述什么内容？

略读下面短文，并回答问题。

Mt. Mckinley's eternal snowcap towers 20,320 feet—the highest point on the North American continent. Wide valleys, worn by such meandering streams as the Kuskokwim and the mighty Yukon, are filled

Unit Four   History of Science and Technology in Ancient China

with unique and colorful plants and animals. Big game hunting and fishing are unparalleled. Moose, bear, Dall sheep and caribou are plentiful, Arctic grayling, salmon, and spectacular trout abound in lakes and rivers. About 95 percent of Alaska is still public domain, where adventure can be enjoyed on a grand scale. Alaska, home of the Eskimo, Indian and Aleut, remains untamed.

(1) Which is the topic sentence?

    A) The first sentence.   B) The third sentence.

    C) The last sentence.   D) It is implied.

(2) Which outline illustrates the structure of the paragraph?

    A) ▽   B) △   C) ◇   D) ▭

(3) Which sentence best expresses the main idea?

    A) Big game hunting and fishing are Alaska's primary tourist attractions.

    B) Traveling through Alaska is difficult since much of it is untamed.

    C) Alaska, a vast wilderness, offers thrilling opportunities for adventure.

    D) Alaska, home of moose, bear, and caribou, is noted for its lakes and rivers.

(4) The paragraph could be entitled _____.

    A) Migrations of Arctic Animals   B) Big Game Hunting and Fishing

    C) The Untamed Giant   D) Arctic Explorations

(5) The main idea of the paragraph suggests that Alaska would appeal to _____.

    A) thrill seekers   B) skiers

    C) tourists   D) mountain climbers

分析：

    全文六句展现的是阿拉斯加一幅原始、荒野、人烟稀少、杂草丛生、野兽横行的画面。从句意的概括性和句子的结构不难看出文章最后一句是主题句，前四句是细节句，第五句是主题句的铺垫。故题 1 答案是 C。

    由此也可推知本文的逻辑结构呈△形，故题 2 答案为 B。

    第 3 题只有 C 与主题句意思最近，并为每个细节句所支撑。

    题 4 答案为 C，运用了修辞上的暗喻，把 Alaska 比作未驯服的巨人，符合文章标题简洁、有趣引人的特点，且与主题句中的 untamed 相吻合。

题 5 根据文章主题思想（题 3），阿拉斯加当然对冒险家最有吸引力，故题 5 答案为 A。

在确立一篇文章的中心思想时，我们也可以先找出各段的主题句，根据这些主题句再结合文章的一些相关信息、重要细节和关键词，然后确立文章的中心思想。

从本讲座中罗列的大量例子可以看出，中心思想是贯穿全文的主线，体现了段落和语篇的方向，对段落和语篇起着指导和限定作用。段落和语篇的中心思想可以通过篇章结构、主题、主题句、标题和作者或文章的意图来体现。

Read the following passages and answer the questions.

1

Doctors are of the opinion that most people cannot live beyond 100 years; but a growing number of scientists believe that the aging process can be controlled. There are more than 12,000 Americans over 100 years old who receive Social Security benefits, and their numbers are increasing each year. Dr. James Langley of Chicago claims that, theoretically and under ideal conditions, animals including man can live six times longer than their normal period of growth. A person's period of growth lasts approximately 25 years. If Dr. Langley's theory is accurate, generations can expect a life span of 150 years.

(1) Which sentence best expresses the main idea?

A) Within a few generations, most people will probably live for 150 years.

B) Social Security pensioners are steadily increasing in number.

Unit Four History of Science and Technology in Ancient China

   C) Physicians and scientists disagree regarding man's possible life span.

   D) Man's normal period of growth compares with that of animals.

(2) The main idea of the paragraph deals with _____.

   A) expected longevity   B) retirement benefits

   C) population control   D) impractical theories

(3) The author develops his main idea by _____.

   A) describing the conditions necessary for a long life

   B) explaining the findings of an expert

   C) comparing man with other animals

   D) quotinga popular medical opinion

(4) Underline the topic sentence.

2

Recently, scientists have become increasingly concerned about pollution and its effects on weather. Actually, man's pollution of the atmosphere is not nearly as great as nature's: In fact, man is the lesser offender compared to nature itself. Specifically, active volcanoes, evaporation of impure water, and wind erosion of the landscape spew 8.5 billion tons of chemicals into space each year. Man's contribution consists of a mere half billion tons. What man contributes, though, is unable to escape very far into the atmosphere, so he must breathe, eat, and drink his own pollutants. Further research by environmental scientists is required to determine if gradual weather changes are related to atmospheric pollution.

(1) Which sentence best expresses the main idea?

   A) Man's pollution is less than nature's but is more harmful.

B) Nature's pollution affects man directly but accounts for only a fraction of the total pollution.

C) Man and nature pollute the atmosphere and directly influence the weather.

D) Man's pollution is less than nature's and is of minor concern.

(2) Which of the following idea does the paragraph support?

A) Steps should be taken to control the weather.

B) Nature's pollution can be controlled.

C) Man is committed to a cleaner environment.

D) Study is needed into the causes of weather changes.

(3) One of the aims of the paragraph is _____.

A) to persuade the reader to stop polluting

B) to inform the reader of recent research

C) to present facts and opinions about pollution

D) to argue in favor of government controls of pollution

(4) Underline the sentence which supports the correct answer for Question 2.

Part B Reading

Science and Technology of the Tang Dynasty

The history of science and technology in China is both long and rich with many contributions to science and technology. In antiquity, independently of other civilizations, ancient Chinese philosophers made significant advances in science, technology, mathematics, and astronomy. Traditional Chinese medicine, acupuncture and herbal medicine were also practiced. Among the earliest inventions were the abacus, the "shadow clock", and the first items such as Kongming lanterns. The Four Great Inventions: the compass, gunpowder, papermaking, and printing, were among the most important technological advances, only known in Europe by the end of the Middle Ages. The Tang Dynasty (618–907 A.D.) in particular was a time of great innovation.

The Tang Dynasty (618 C.E.–907 C.E.) of ancient China witnessed many advancements in Chinese science and technology, with various developments in woodblock printing, timekeeping, mechanical engineering, medicine, structural engineering, cartography, and alchemy.

Woodblock printing

Woodblock printing made the written word available to vastly greater audiences. One of the world's oldest surviving printed documents is a miniature Buddhist *dharani*[1] sutra[2] unearthed at Xi'an (then called Chang'an, the capital of the Tang Dynasty) in 1974 and dated roughly from 650 to 670. The *Diamond Sutra*[3] is the first full-length book printed at regular size, complete with illustrations embedded with the text and dated precisely to 868. Among the earliest documents to be printed were Buddhist texts as well as calendars, the latter essential for calculating and marking which days were auspicious and which days were not. Although the later

Bi Sheng's [4] movable type printing in the 11th century was innovative for his period, woodblock printing that became widespread in the Tang would remain the dominant printing type in China until the more advanced printing press from Europe became widely accepted and used in East Asia. The first use of the playing card during the Tang Dynasty was an auxiliary invention of the new age of printing.

Clockworks and timekeeping

Technology during the Tang period was built also upon the precedents of the past. The mechanical gear systems of Zhang Heng [5] (78–139) and Ma Jun [6] (fl. 3rd century) gave the Tang engineer, astronomer, and monk Yi Xing [7] (683–727) inspiration when he invented the world's first clockwork escapement [8] mechanism in 725. This was used alongside a clepsydra clock and waterwheel to power a rotating armillary sphere in representation of astronomical observation. Yi Xing's device also had a mechanically timed bell that was struck automatically every hour, and a drum that was struck automatically every quarter hour; essentially, a striking clock. Yi Xing's astronomical clock and water-powered armillary sphere became well known throughout the country, since students attempting to pass the imperial examinations by 730 had to write an essay on the device as an exam requirement. However, the most common type of public and palace timekeeping device was the inflow clepsydra. Its design was improved c. 610 by the Sui-Dynasty engineers Geng Xun and Yuwen Kai. They provided a steelyard balance [9] that allowed seasonal adjustment in the pressure head of the compensating tank and could then control the rate of flow for different lengths of day and night.

Mechanical delights and automatons

There were many other mechanical inventions during the Tang era. This included a three-feet-tall mechanical wine server of the early 8th century that was in the shape of an artificial mountain, carved out of iron and rested on a lacquered-wooden tortoise frame. This intricate device used a hydraulic pump that siphoned wine out of metal dragon-headed faucets,

Unit Four History of Science and Technology in Ancient China

as well as tilting bowls that were timed to dip wine down, by force of gravity when filled, into an artificial lake that had intricate iron leaves popping up as trays for placing party treats.

Although the use of a teasing mechanical puppet in this wine-serving device was certainly ingenious, the use of mechanical puppets in China date back to the Qin Dynasty [10] (221–207 B.C.) while Ma Jun in the 3rd century had an entire mechanical puppet theater operated by the rotation of a waterwheel. There are many stories of automatons used in the Tang, including general Yang Wulian's wooden statue of a monk who stretched his hands out to collect contributions; when the amount of coins reached a certain weight, the mechanical figure moved his arms to deposit them in a satchel. This weight-and-lever mechanism was exactly like Heron's penny slot machine [11]. Other devices included one by Wang Ju, whose "wooden otter" could allegedly catch fish; Needham suspects a spring trap of some kind was employed here.

Medicine

The Chinese of the Tang era were also very interested in the benefits of officially classifying all of the medicines used in pharmacology. In 657, Emperor Gaozong of Tang [12] (r. 649–683) commissioned the literary project of publishing an official *materia medica*, complete with text and illustrated drawings for 833 different medicinal substances taken from different stones, minerals, metals, plants, herbs, animals, vegetables, fruits, and cereal crops. In addition to compiling pharmacopeias, the Tang fostered learning in medicine by upholding imperial medical colleges, state examinations for doctors, and publishing forensic manuals for physicians. Authors of medicine in the Tang include Zhen Qian (d. 643) and Sun Simiao [13] (581–682), the former who first identified in writing that patients with diabetes had an excess of sugar in their urine, and the latter who was the first to recognize that diabetic patients should avoid consuming alcohol and starchy foods. As written by Zhen Qian and others in the Tang, the thyroid glands of sheep and pigs were successfully used to treat goiters; thyroid extracts were not used to treat patients with goiter in the West until 1890.

Structural engineering

In the realm of technical Chinese architecture, there were also government standard building codes, outlined in the early Tang book of the *Yingshan Ling* (National Building Law). Fragments of this book have survived in the *Tang Lü* (The Tang Code), while the Song Dynasty architectural manual of the *Yingzao Fashi*[14] (State Building Standards) by Li Jie (1065–1101) in 1103 is the oldest existing technical treatise on Chinese architecture that has survived in full. During the reign of Emperor Xuanzong of Tang[15] (712–756) there were 34,850 registered craftsmen serving the state, managed by the Agency of Palace Buildings (Jingzuo Jian).

Cartography

In the realm of cartography, there were further advances beyond the map-makers of the Han Dynasty. When the Tang chancellor Pei Ju[16] (547–627) was working for the Sui Dynasty as a Commercial Commissioner in 605, he created a well-known gridded map with a graduated scale in the tradition of Pei Xiu[17] (224–271). The Tang chancellor Xu Jingzong[18] (592–672) was also known for his map of China drawn in the year 658. In the year 785 the Emperor Dezong[19] had the geographer and cartographer Jia Dan[20] (730–805) complete a map of China and her former colonies in Central Asia. Upon its completion in 801, the map was 9.1 m (30 ft) in length and 10 m (33 ft) in height, mapped out on a grid scale of one inch equaling one hundred *li* (Chinese unit of measuring distance). A Chinese map of 1137 is similar in complexity to the one made by Jia Dan, carved on a stone stele with a grid scale of 100 *li*. However, the only type of map that has survived from the Tang period is star charts[21]. Despite this, the earliest extant terrain maps of China come from the ancient State of Qin; maps from the 4th century B.C. were excavated in 1986.

Alchemy, gas cylinders, and air conditioning

The Chinese of the Tang period employed complex chemical formulas

Unit Four   History of Science and Technology in Ancient China

for an array of different purposes, often found through experiments of alchemy. These included a waterproof and dust-repelling cream or varnish for clothes and weapons, fireproof cement for glass and porcelain wares, a waterproof cream applied to silk clothes of underwater divers, a cream designated for polishing bronze mirrors, and many other useful formulas. The vitrified, translucent ceramic known as porcelain was invented in China during the Tang, although many types of glazed ceramics preceded it.

Ever since the Han Dynasty (202 B.C.–220 A.D.), the Chinese had drilled deep boreholes to transport natural gas from bamboo pipelines to stoves where cast iron evaporation pans boiled brine to extract salt. During the Tang Dynasty, a gazetteer of Sichuan province stated that at one of these 182 m (600 ft) "fire wells", men collected natural gas into portable bamboo tubes which could be carried around for dozens of km and still produce a flame. These were essentially the first gas cylinders; Robert Temple assumes some sort of tap was used for this device.

The inventor Ding Huan (fl. 180 A.D.) of the Han Dynasty invented a rotary fan for air conditioning, with seven wheels 3 m (10 ft) in diameter and manually powered. In 747, Emperor Xuanzong had a "Cool Hall" built in the imperial palace, which the *Tang Yulin* [22] describes as having water-powered fan wheels for air conditioning as well as rising jet streams of water from fountains. During the subsequent Song Dynasty, written sources mentioned the air conditioning rotary fan as even more widely used.

New Words and Expressions

acupuncture 针灸
herbal medicine 中草药
abacus 算盘
woodblock [印] 木版；木刻（画）
timekeeping （精确）计时, 时间测量

cartography 地图制作
miniature 小型的
Buddhist 佛教的
embedded 嵌入的
auspicious 吉利的，吉祥的

auxiliary 辅助的；附属的
clockwork 时钟机构
precedent 先例，前例；惯例
escapement 擒纵机构
clepsydra 漏壶（一种古时用的计时器）
armillary 圆环的
inflow 内流，进水
steelyard 提秤
automaton [计] 机器人；自动机械装置
lacquered 上漆的，涂漆的，喷漆的
intricate 复杂的；错综的
siphon 用虹吸管吸出（或输送）；抽取
faucet 龙头，旋塞
tilt 使倾斜
dip 吊下来；下降；下垂
gravity 重力
pop 弹跳，弹起
treat 难得的乐事；款待；请客
rotation 旋转
contribution 捐款
deposit 存放
satchel 书包，小背包
otter 水獭
allegedly 据称
pharmacology 药理学，药物学
commission 委任，任命
materia medica [拉][总称]药物；药物学（论著）
pharmacopeia 药典
foster 培养，鼓励
uphold 支持；拥护
forensic 辩论的
diabetes 糖尿病
urine 尿
starchy 淀粉的；浆糊的
thyroid glands 甲状腺
goiter [医] 甲状腺肿
code 法典，法规
chancellor 大臣
grid 地图的坐标方格
stele 石碑，石柱
extant （书籍、文件、画等）现存的；尚存的
vitrified 陶瓷的
translucent 半透明的
ceramic 陶器的
porcelain 瓷器
glaze 上釉于，上光于
borehole 地上凿孔
evaporation 蒸发；发散
brine 盐水，海水
gazetteer 地名词典，地名索引
rotary 旋转的，轮转的，转动的

Unit Four　History of Science and Technology in Ancient China

📝 Notes

1. *dharani* 陀罗尼（佛教用语，源自古印度语），汉译后经常出现于佛教经典中。

2. *sutra* 契经（又称线经，简称经，音译修多罗）。原义为线，是印度教和佛教的一种文体。

3. *Diamond Sutra*《金刚般若波罗蜜经》（简称《金刚经》）是大乘佛教重要经典之一，为出家、在家佛教徒常所颂持。20 世纪初出土于敦煌的《金刚经》为世界最早的雕版印刷品之一，现存于大英图书馆。

4. Bi Sheng（约 970—1051）毕昇，中国发明家，发明了活字版印刷术。

5. Zhang Heng（78—139）张衡，中国东汉天文学家、数学家、发明家、地理学家、制图学家和诗人等。张衡发明了世界第一个水力浑天仪，能探测到 500 公里外的地震。张衡被认为是通才。1802 号小行星以他的名字命名。

6. Ma Jun 马钧，三国时期著名的发明家。因为在传动机械方面造诣很深，有"天下之名巧"的美誉。善于发明机械、武器。

7. Yi Xing（683—727）一行禅师，唐代僧人和天文学家，被唐人呼为"一公"。著有《大日经疏》《大衍历》等书。

8. escapement 擒纵器是使用于钟表机械的零件。擒纵机构最早由希腊的拜占廷人斐罗（Philo of Byzantium）所发明，当时用于机械式盥洗台的一部分，然而用于钟表机械的擒纵机制最早是由唐代比丘僧和一行开发。

9. steelyard balance 秤是传统的量重工具，其主体为木或竹竿，一端有一大钩，另一端挂有重物（秤砣）。将要量的东西挂在钩上，移动秤砣利用杠杆原理使两边达至平衡并读取读数。

10. The Qin Dynasty（前 221—前 207）秦朝是由战国时代后期的秦国发展起来的统一大帝国。秦朝开国君主秦王政自称始皇帝，从此有了皇帝这一词语。国号"秦"，王室嬴姓，故史书上别称"嬴秦"，以别于其他国号为秦的政权。秦朝从统一六国到灭亡，只有 15 年。

11. slot machine 老虎机（角子机或全名吃角子老虎机）是一种经常在赌场见到的赌博机器。

12. Emperor Gaozong of Tang（628—683）唐高宗李治，唐朝第三任皇帝，唐太宗李世民第九子、嫡三子。

13. Sun Simiao（581—682）孙思邈，唐朝著名的医师与道士。他是中国乃至世界史上著名的医学家和药物学家，被誉为"药王"，许多华人奉之为"医神"。他著有《备急千金要方》（简称《千金要方》），该书已接近现代临床医学的分类方法。《千金翼方》是对《千金要方》的补编。他将《伤寒论》的内容较完整地收集在《千金翼方》之中。

14. *Yingzao Fashi*《营造法式》是中国第一本详细论述建筑工程做法的官方著作。此书于北宋元符三年（1100年）编成，由李诫所作。书中规范了各种建筑做法，详细规定了各种建筑施工设计、用料、结构、比例等方面的要求。

15. Emperor Xuanzong of Tang（685—762）唐玄宗李隆基（712—756年在位），被称为唐明皇。

16. Pei Ju（547—627）裴矩，隋末唐初政治家。其一生中最重要的业绩是为隋炀帝经营西域，大力开发商业，引胡商前往长安、洛阳，以首都贸易取代边境贸易。与此同时，裴矩收集了西域各国山川地理、人物风俗等资料，绘画各国贵族庶人的服饰仪表，制成《西域图记》三卷，并绘造地图，记录各地险要，献给隋炀帝。所撰《西域图记》现仅存书序，记载了从敦煌到西海（今地中海）的三条主要路线，是中西交通的重要史料。还著有《开业平陈记》《高丽风俗》等书。

17. Pei Xiu（224—271）裴秀，西晋政治家，地图学家。他著有中国最早有记载的地图集《禹贡地域图》，并在其序中提出了对中国古代地图绘制影响很大的"制图六体"。在明末清初欧洲的地图投影方法传入中国之前，裴秀的"六体"一直是中国古代绘制地图的重要原则，对于中国传统地图学的发展影响极大。

18. Xu Jingzong（592—672）许敬宗，著有《文馆词林》。

19. Emperor Dezong（742—805）唐德宗李适。

20. Jia Dan（730—805）贾耽，唐朝官员，博学好古，尤以精通地理学著称于世。

21. star charts 星图，不同于传统地理图集或者天体照片。现代星图是把

Unit Four History of Science and Technology in Ancient China

夜空中持久的特征精确描述或绘制，古时的星图（如敦煌星图）最初只以小圆圈或一样的圆点附以连线表示星官与星座，后期才陆续加上标示黄道、银道等参考线。

22. *Tang Yulin*《唐语林》。

Exercises

I. Find out the English equivalents to the following Chinese terms from the passage.

1. 刻版印刷
2. 结构工程
3. 地图制作
4. 金刚经
5. 佛经
6. 活字印刷
7. 浑天仪
8. 自动定时钟
9. 天文钟
10.（中国封建社会的）科举
11. 计时装置
12. 水力泵
13. 药性物质
14. 有坐标网格的地图
15. 星图
16. 天然气
17. 气瓶
18. 空气（温度）调节
19. 瓷器（件）
20. 旋转的风扇

II. Choose the best answer according to the information in the passage.

1. Which book was the first one that was printed at regular size with illustrations?

    A) *Dharani sutra*.
    B) *Diamond Sutra*.
    C) *The Books of Poetry*.
    D) None of the above.

2. Who invented the world's first clockwork escapement mechanism?

    A) Monk Yi Xing
    B) Zhang Heng
    C) Ma Jun
    D) Geng Xun

3. What was the mechanical wine server made of in the Tang Dynasty?

    A) Steel.　　　B) Wood.　　　C) Silver.　　　D) Iron.

4. In which dynasty, was there a big step in the realm of cartography?

    A) The Sui Dynasty.　　　　B) The Tang Dynasty.

    C) The Han Dynasty.　　　　D) The Qing Dynasty.

5. What kind of power was it used to drive the air-conditioning in the Han Dynasty?

    A) Water-powered.　　　　B) Wind-powered.

    C) Gas-powered.　　　　　D) Steam-powered.

Part C Extended Reading

Science in Ancient China

Mathematics

From the earliest the Chinese used a positional decimal system on counting boards in order to calculate. To express 10, a single rod is placed in the second box from the right. The spoken language uses a similar system to English: e.g. four thousand two hundred seven. No symbol was used for zero. By the first century B.C., negative numbers and decimal fractions were in use and *The Nine Chapters on the Mathematical Art* [1] included methods for extracting higher order roots by Horner's method [2] and solving linear equations and by Pythagoras' theorem [3]. Cubic equations were solved in the Tang Dynasty and solutions of equations of order higher than 3 appeared in print in 1245 by Ch'in Chiu-shao [4]. Pascal's triangle [5] for binomial coefficients was described around 1100 by Jia Xian [6].

Although the first attempts at an axiomatization of geometry appear in the Mohist [7] canon in 330 B.C., Liu Hui [8] developed algebraic methods in geometry in the 3rd century and also calculated pi to 5 significant figures. In 480, Zu Chongzhi [9] improved this by discovering the ratio $\frac{355}{113}$ which remained the most accurate value for 1200 years.

Astronomy

Astronomical observations from China constitute the longest continuous sequence from any civilization and include records of sunspots (112 records from 364 B.C.), supernovas (1054), lunar and solar eclipses. By the 12th century, they could reasonably accurately make predictions of eclipses, but the knowledge of this was lost during the Ming Dynasty, so that the Jesuit Matteo Ricci [10] gained much favor in 1601 by his predictions. By 635 Chinese astronomers had observed that the tails of comets always point away from the sun.

From antiquity, the Chinese used an equatorial system for describing the skies and a star map from 940 was drawn using a cylindrical(Mercator) projection. The use of an armillary sphere is recorded from the 4th century B.C. and a sphere permanently mounted in equatorial axis from 52 B.C. In 125 A.D. Zhang Heng used water power to rotate the sphere in real time. This included rings for the meridian and ecliptic. By 1270 they had incorporated the principles of the Arab torquetum.

Seismology (the scientific study of earthquake): To better prepare for calamities, Zhang Heng invented a seismometer in 132 C.E. which provided instant alert to authorities in the capital Luoyang that an earthquake had occurred in a location indicated by a specific cardinal or ordinal direction [11]. Although no tremors could be felt in the capital when Zhang told the court that an earthquake had just occurred in the northwest, a message came soon afterwards that an earthquake had indeed struck 400 km to 500 km northwest of Luoyang (in what is now modern Gansu). Zhang called his device the "instrument for measuring the seasonal winds and the movements of the Earth"(*Houfeng didong yi*[12]), so-named because he and others thought that earthquakes were most likely caused by the enormous compression of trapped air.

There are many notable contributors to the field of Chinese science throughout the ages. One of the best examples would be Shen Kuo[13] (1031–1095), a polymath scientist and statesman who was the first to describe the magnetic-needle compass used for navigation, discovered the concept of true north, improved the design of the astronomical gnomon, armillary sphere, sight tube, and clepsydra, and described the use of drydocks to repair boats. After observing the natural process of the inundation of silt and the find of marine fossils in the Taihang Mountains (hundreds of miles from the Pacific Ocean), Shen Kuo devised a theory of land formation, or geomorphology. He also adopted a theory of gradual climate change in regions over time, after observing petrified bamboo found underground at

Unit Four   History of Science and Technology in Ancient China

Yan'an, Shaanxi province. If not for Shen Kuo's writing, the architectural works of Yu Hao [14] would be little known, along with the inventor of movable type printing, Bi Sheng (990–1051). Shen's contemporary Su Song [15] (1020–1101) was also a brilliant polymath, an astronomer who created a celestial atlas of star maps, wrote a pharmaceutical treatise with related subjects of botany, zoology, mineralogy, and metallurgy, and had erected a large astronomical clocktower in Kaifeng city in 1088. To operate the crowning armillary sphere, his clocktower featured an escapement mechanism and the world's oldest known use of an endless power-transmitting chain drive.

New Words and Expressions

Mohist 墨家；墨子信徒	meridian 子午线
canon 学说	ecliptic 黄道
algebraic 代数的，代数学的，代数上的	seismology 地震学
	calamity 灾难
sunspot 太阳黑子	polymath 博学的；博学之人
supernova 超新星	gnomon 时针
Jesuit 耶稣会	drydock 干船坞
equatorial 赤道的，赤道附近的	inundation 泛滥
cylindrical (Mercator) projection 圆柱（麦卡托）投影	silt 淤泥
	petrified 石化的
mount 安装	

Notes

1. *The Nine Chapters on the Mathematical Art*《九章算术》，现存最早的中国古代数学著作之一，它是《算经十书》中最重要的一种。其作者已不可考。一般认为它是经历代各家的增补修订，而逐渐成为现今定本。《九章算术》总结了自周朝以来的中国古代数学，它既包含了以前已经解决了的数学问题，又有汉朝时新发现的数学成就。一般认为，它标志着中国古代数学体系的形成，是中国古代数学体系的初期代表作。

2. Horner's method 秦九韶算法或霍纳算法。秦九韶算法是中国南宋时期的数学家秦九韶表述求解一元高次多项式方程的数值解的算法——正负开方术。19 世纪初，英国数学家威廉·乔治·霍纳重新发现并证明，后世称作霍纳算法。

3. Pythagoras' theorem 勾股定理，又称毕达哥拉斯定理。它是一个基本的几何定理，传统上认为是由古希腊的毕达哥拉斯所证明。勾股定理在中国数学史上源远流长，是中算的重中之重。《周髀算经》记载了勾股定理的特例（勾三股四弦五），相传是在公元前 11 世纪商代由商高发现，故又有称之为商高定理；《九章算术》第九卷《句股》详细讨论了勾股定理的运用，魏国数学家刘徽反复运用勾股定理求圆周率，他的《海岛算经》更进一步将勾股理论发展成为领先世界一千余年的四次勾股重差测量术。

4. Ch'in Chiu-shao（1208—1261）秦九韶，中国南宋数学家。著作有《数书九章》，其中的大衍求一术（一次同余方程组问题的解法，现称中国剩余定理）和秦九韶算法（高次方程正根的数值求法）是具有世界意义的重要贡献。

5. Pascal's triangle 杨辉三角形，又称贾宪三角形、帕斯卡三角形，是二项式系数在三角形中的一种几何排列。

6. Jia Xian 贾宪，11 世纪前半叶中国北宋数学家。贾宪对中国古代数学有极其重要的贡献，许多宋元的数学成就都起源于贾宪。

7. Mohist 墨家，为中国古代春秋战国时期的诸子百家之一，创始人为墨翟，世称墨子，墨家之名从创始人而得。之后由于西汉汉武帝的独尊儒术政策、社会心态的变化以及墨家本身并非人人可达的艰苦训练、严厉规则及高尚思想，墨家在汉武帝之后基本消失。

8. Liu Hui 刘徽，三国时代魏国数学家。三国魏景元四年（263）注《九章算术》（九卷），后撰《重差》，作为《九章算术注》的第十卷。

9. Zu Chongzhi（429—500）祖冲之，刘宋时代数学家、天文学家。他的主要成就在数学、天文历法和机械制造三个领域。此外历史记载祖冲之精通音律，擅长下棋，还写有小说《述异记》。祖冲之著述很多，但大多都已失传。遗留下来的祖冲之的数学贡献主要有他对圆周率的计算结果和球体体积

Unit Four　History of Science and Technology in Ancient China

的计算公式。

10. Matteo Ricci（1552—1610）利玛窦，意大利耶稣会传教士和学者。

11. cardinal or ordinal direction 基本方位和居间方位。（罗盘上的）基本方位指北、南、东、西，居间方位指东北、东南、西南、西北。

12. *Houfeng didong yi* 候风地动仪。

13. Shen Kuo（1031—1095）沈括。在物理学、数学、天文学、地学、生物医学等方面都有重要的成就和贡献，在化学、工程技术等方面也有相当的成就。此外，沈括在文学、音乐、艺术、史学等方面都有一定的造诣。而他突出的成就主要集中在《梦溪笔谈》中。《宋史·沈括传》称他"博学善文，于天文、方志、律历、音乐、医药、卜算无所不通，皆有所论著"。英国科学史家李约瑟评价沈括是"中国科学史上的坐标"。

14. Yu Hao 喻皓，又作预浩、喻浩、预皓，中国古代有名建筑工匠，五代末北宋初浙江杭州一带人，他是一位出身卑微的建筑工匠。他的生卒年代历史上缺乏记载，只知道他在北宋初年当过都料匠（掌管设计、施工的木工），长期从事建筑工作，曾主持汴梁（今开封）开宝寺木塔的建造。著有《木经》三卷，惜已失传，仅在沈括著《梦溪笔谈》中有简略记载。欧阳修《归田录》称赞他是"国朝以来木工一人而已"。他善于建筑多层的宝塔和楼阁。

15. Su Song（1020—1101）苏颂，北宋天文学家和药物学家。其祖先在唐末随王潮入闽，世代为闽南望族。苏颂出身于书香门第，从小聪敏好学，接受了严格的家庭教育。

📝 Exercises

I. **Find out the English equivalents to the following Chinese terms from the passage.**

1. 十进制
2. 霍纳算法 / 秦九韶算法
3. 线性方程式
4. 勾股定理
5. 三项方程
6. 二项式系数
7. 帕斯卡斯三角形 / 杨辉三角形
8. 赤道轴
9. 地震仪
10. 观察管

11. 海洋生物化石　　　12. 天体图集

13. 医药论文　　　　14. 地貌学

II. Answer the following questions by using the sentences in the text or in your own words.

1. What are the achievements on mathematics in ancient China?

2. What are included in the astronomical records in ancient China?

3. Did Zhang Heng forecast the earthquake in northwest of Luoyang precisely?

4. What are main contributions of Shen Kuo in Chinese science? Please give three examples.

Unit Five

Science and Technology in the Middle Ages and Renaissance

Part A　Lecture

辨认事实与细节

上一单元的讲座我们讨论了如何确定作者要表达的中心思想，本单元我们将继续探讨作者是用哪些事实和细节来阐述和发展中心思想的。辨认事实与细节不是阅读中最重要的技巧，捕捉主题、概括推论等阅读技巧无疑要比其更关键、更困难。但为了有效地捕获文章主题，基于短文内容进行概括推论，我们必须能够辨认事实与细节。因此，在掌握文章（或段落）中心思想的基础上，我们还应学会抓住用来阐述和发展中心思想的主要事实，或者按要求找出特定细节。这是通过阅读获得信息的重要能力；掌握了这种能力对我们提高阅读效率有很大的帮助。

1　辨认事实与细节的阅读技巧

在文章中，作者总是要通过一系列的具体内容来支持、论证、阐述、说明和展开中心思想的。阅读过程中，我们首先要明白文章中有哪些具体内容，即 who、what、when、where、why。试分析下例。

  Japan is an island country in the Pacific Ocean. There are four main islands and more than 3000 small ones that stretch from north to south for about 1300 miles. As for the climate of the country, it is hot and humid in the summer; cold and wet in the winter.

  The population of Japan is about 110 ½ million. The average population density of the country is about 678 persons per square mile. Just in comparison, the average population density of the United States was listed at fifty-seven persons per square mile. About two thirds of the entire population live in Japan's cities. The other one third lives in

Unit Five   Science and Technology in the Middle Ages and Renaissance

the suburbs or in the rural areas. As a matter of fact, no other country in southern or eastern Asia has such a large urban, or city, population.

Japan has a wide variety of resources, but their quantity is really very small. So Japan must import most of its minerals. Japan must also import large quantities of food and raw materials. And in order to pay for these goods, the Japanese export products that they manufacture, such as TVs, cars, and so forth. Since World War II, Japan has become one of the world's chief industrial and manufacturing countries, and the Japanese people enjoy a high standard of living.

As many people know, respect for learning and tradition and a love of beauty are traditional Japanese characteristics. The Japanese often express this love of beauty in the creation and enjoyment of beautiful art, sculpture, ceramics, music, and theater productions.

Although they are very industrious, hard-working, and enterprising people, the Japanese still manage to enjoy their leisure time. They really enjoy sports, such as soccer, swimming, judo, wrestling, Karate, and baseball. They also like to take vacation trips in their own country and abroad.

(1) How many main islands make up the country of Japan?

 A) Five. B) Four. C) Seven. D) Six.

(2) How many miles long is the country of Japan from north to south?

 A) Three hundred.

 B) One thousand.

 C) One thousand three hundred.

 D) Three thousand one hundred.

(3) How many people per square mile are there in Japan?

 A) 578. B) 778. C) 621. D) 678.

(4) What percentage of Japanese population live in the rural areas?

 A) 1/3. B) 2/3. C) 1/4. D) 2/4.

(5) What must Japan import from abroad in large quantities?

 A) Food and raw materials. B) Materials and products.

 C) Food and goods. D) TVs and cars.

(6) Since when has Japan become a chief industrial and manufacturing country?

 A) World War I. B) 1963. C) World War II. D) 1950.

(7) What attitude do the Japanese take for learning?

 A) Love. B) Respect. C) Dislike. D) Hatred.

分析：从问题中可知，本篇短文主要考核文章的具体内容。阅读短文后，我们很容易找到答案。第1—7题答案分别为 B、C、D、A、A、C、B。

在阅读过程中，我们除了要明白文章中有哪些具体内容外，同时，还要辨认事实与细节，即分清主次信息。如果分不清主次信息，就可能对作者要表达的中心思想的理解出现错误。请分析下例。

  In 1983, the National Commission on Excellence in Education said that "a tide of mediocrity" had swept over public schools. As a result, it suggested many reforms. It asked for required classes in "the new basics". This would mean four years of English for everyone. All students would also need three years of math, science and social studies and a half year of computer science. Two years of a foreign language were recommended for those students planning to go on to college. Besides "the new basics", the commission also asked for seven-hour school days and a longer school year. In addition, it suggested better pay for teachers and higher standards for would-be teachers.

分析：在辨认事实与细节时要注意区分主要信息和次要信息。主要信息是文章中对于主题主旨具有典型意义的重要事实，是中心思想的基础，是文章的基本框架。次要信息就像大树上的细枝，是构成主要信息的局部因素或充实事实的例子、数据等细微末节。

  显然，上文中的主要信息是第1、2句用一因一果指出的改革（reforms）。因为所有公共学校（public school）的不景气（mediocrity），所以 the National Commission 提出改革。接着下面各句陈述其各项建议细节（即次要信息），用 also、besides、in addition 等词语把各个细节（即次要信息）连成一体去充实主要信息。

从上例可以看出区分主次信息的重要性。那么，在阅读过程中怎样辨认

Unit Five　Science and Technology in the Middle Ages and Renaissance

主要信息和次要信息呢？以下我们从段落层次的规律、组织文章思想的逻辑方式和与中心思想的关系等角度进行探讨。

1·1 通过段落层次的规律分清信息的主次

从文章段落组织方式来看，段落组成是有层次的。主题句为第一层次，概括性最强；直接阐述主题句的为第二层次；直接阐述第二层次的为第三层次。以此类推，其概括性逐渐减弱，具体性逐渐加强。段落层次结构见图5-1。

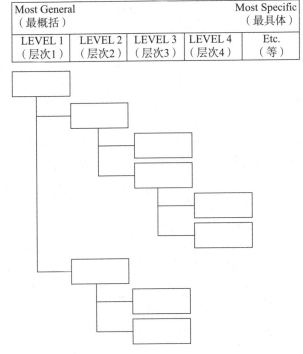

图 5-1　段落层次结构框图

试分析下面例子的段落层次。

例 3

(1) Vilma likes traveling by bus better than by airplane for four reasons. (2) First, it costs less. (3) She can ride by bus from New York to Chicago for

$ 20. (4) Yet, the air fare between the two cities is $ 78. (5) Second, traveling by bus gives her a closer look at the cities and countryside than she could get from a plane. (6) For example, she enjoys driving through the big cities of Philadelphia, Pittsburgh, and Fort Wayne. (7) Third, Vilma finds that the passengers on the bus are often closer to her own age. (8) So, they are easy to talk with on the shared travel adventure. (9) In contrast, many airline passengers are businessmen who keep busy with their work while flying. (10) Fourth, riding a bus allows Vilma to explore any stop along the way. (11) For instance, once she got off the bus at a small town in Indiana and spent the night with a girl friend. (12) Then, the next day she took another bus to Chicago. (13) For these reasons, Vilma usually prefers buses instead of planes unless she has to get somewhere in a hurry.

分析：该段落可分为三个层次（段落层次见图 5-2），第一层次为中心思想，第二层次为主要信息，第三层次为次要信息。第二层次（四条理由）是直接阐述层次一（中心思想）的，与主题的关系最密切。第三层次共 7 句，分别进一步阐述了第二层次中提出的理由：句 3 与句 4 是支持理由一（句 2）的具体事实，句 6 是支持理由二（句 5）的具体事实，句 8 与句 9 是支持理由三（句 7）的具体事实，句 11 与句 12 是支持理由四（句 10）的具体事实。第三层次也是服务于阐述主题（层次一）的，但它们与层次一的关系是间接的。由此看来，分清了段落的层次，也就看清了作者的思路，因而便于分清信息的主次。当然在同一层次内列举的每个事实的重要性有时也并不完全等同，也有主次轻重之分。且看下节的分析。

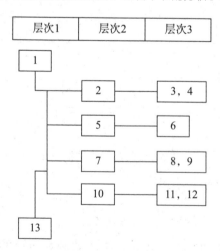

图 5-2 段落的三个层次图

Unit Five　Science and Technology in the Middle Ages and Renaissance

1·2 通过组织文章思想的逻辑方式分清信息的主次

一般来说，作者是按一定的逻辑方式组织文章的思想的。在阅读中，如能从作者写作的角度分析文章，就能比较容易地识别文章主次信息的组织方式。

作者组织文章思想的逻辑方式通常有以下几种：

（1）简单排列（simple listing）

作者所列举的各细节彼此间没有特殊的联系，唯一将它们联系在一起的是短文主题那根线。

（2）重要性排序（order of importance）

最重要的细节置于最前，较次要的随后，最次要的最后。或以完全相反的顺序排列，即先提出次要细节，然后层层递进，最后才是最重要的细节。

（3）时间顺序（time order）

细节的出现先后与其实际发生的时间次序相吻合。这种方法更多地用于故事叙述或人物生平简介。

（4）空间顺序（spatial development）

各细节按处所位置的前后左右、东西南北、上下远近有规律地排列。

（5）比较对照（compare and contrast）

两个（或更多的）人、物之间的相似之处放在一起进行比较，相异之处放在一起形成对照。比较和对照经常同时用于一篇文章阐述某些人、物之间的共性和个性。

（6）因果关系（cause and consequence）

一个细节为另一细节的起因或结果时，作者常按事件的因果关系来安排情节，或先因后果，或先果后因。

试分析下列各段中哪些部分是最重要的细节（即主要信息）。

　　Madame Curie and her husband discovered that a piece of pitch-blend produced a darkening of photographic plates out of all proportion to its uranium content. This meant to them that the pitch-blend contained some other element. The Curies decided to try and isolate this new

element. The task took all their time for two years. When they were through, the Curies had discovered not one but two new elements. The first to be discovered was called polonium, after Madame Curie's native Poland; the other element was the famous metal radium.

A) The Curies decided to try and isolate this new element.

B) The task took all their time for two years.

C) When they were through, the Curies had discovered not one but two new elements.

D) The first to be discovered was called polonium, after Madame Curie's native Poland; the other element was the famous metal radium.

分析：在以上四个选项中，D 项为最重要的细节，即主要信息，直接说明沥青混合物里发现的元素是什么。

1·3 通过与中心思想的关系分清信息的主次

我们知道，主要信息是中心思想的基础，它与中心思想有直接的关系，所以，判断是否是主要信息的依据就是看是否围绕中心思想。围绕中心思想、直接阐述和说明中心思想的信息是主要信息，其他的是次要信息。

因此，正确辨认文章（或段落）的主要信息和次要信息，可以从以下几个方面入手：1）找出文章（或段落）的中心思想或主题句；2）找出围绕中心思想的事实；3）弄清事实和中心思想之间的联系；4）找出引导我们发现细节的信号词。

试分析下例，找出阐述或说明中心思想（和主题）的事实。

Have you ever wondered why products come in the colors they do? For instance, why is toothpaste often green or blue and shampoo often golden-yellow? Manufacturers pick the colors that are associated with qualities that consumers value in certain products. For example, it's known that blue symbolizes purity to most people and that green is refreshing. These are both desirable qualities in toothpaste. Manufacturers also know that golden-yellow symbolizes richness (as in real gold or egg yolks), so they frequently choose this color for shampoos and cream rinses—

Unit Five   Science and Technology in the Middle Ages and Renaissance

products in which consumers value richness. Baby products, such as body lotion, are often tinted pink because that is a color commonly associated with softness and gentleness—the very qualities consumers want for their baby's care.

Topic: Colors of Products

Main Idea: Colors are associated with qualities of the products.

Supporting Details:

1) Manufacturers use blue and green colors for toothpaste because customers value them as purity and refreshment.

2) Golden-yellow is chosen for shampoo and cream rinses because it symbolizes richness.

3) Baby products are tinted pink because the color is associated with softness and gentleness needed for their baby's care.

分析：本文讨论的话题是"产品的颜色"，从文中可知，作者要表达的中心思想"产品的质量与颜色相关"来自第 3 句；接着作者使用例证法展开，用了牙膏、洗发水和婴儿乳液三个例子来说明"产品的质量与颜色相关"，并使用了 for example、also 等信号词。

(1) A study of art history might be a good way to learn more about a culture than is possible to learn in general history class. (2) Most typical history courses concentrate on politics, economics, and war. (3) But art history focuses on much more than this because art reflects not only the political values of a people, but also religious beliefs, emotions, and psychology. (4) In addition, information about the daily activities of our ancestors—or of people very different from our own—can be provided by art. (5) In short, art expresses the essential qualities of a time and a place, and a study of it clearly offers us a deeper understanding than can be found in most history books.

(1) Circle the topic sentence from the passage.

　　A) Sentence 1　　　　　B) Sentence 3　　　　　C) Sentence 2

(2) Circle all the sentences with the major details from the passage.

　　A) Sentence 2, 3　　　　B) Sentence 1, 3, 5　　　C) Sentence 3, 4

(3) Circle one of the sentences with the minor details from the passage.

   A) Sentence 3   B) Sentence 2   C) Sentence 5

分析：该短文的第 1 句是主题句，概括了全段的中心内容，关键词语是 a good way to learn more about a culture。句 5 为主题句的重述。句 3、句 4 直接地解释了句首的关键部分，因而是主要事实。句 2 是句 3 的对比句，用于突出句 3，与主题句的关系不如句 3、句 4 直接重要。因此答案为 (1) A; (2) C; (3) B。

2 综合练习

以上我们分析了辨认事实与细节的阅读技巧，知道了通过段落层次的规律、组织文章思想的逻辑方式和与中心思想的关系等手段可以分清主要信息和次要信息，以下我们将对此阅读技巧进行练习。

为了准确辨认事实与细节，我们可以按下列步骤和方法进行：1）确定短文主题或主题句或中心思想；2）确定短文安排细节的结构方式以及组织短文思想的逻辑方式；3）挑选出直接论证主题或主题句或中心思想的主要事实；4）如果是阅读理解题，看问题，并根据问题判定有关信息的大概位置；5）带着问题以查读的方法快速找到该信息的确切位置，再选题验证。

试分析下例。

例 7

The Arctic Circle is an imaginary line about 1650 miles from the North Pole. About one million people live within the area bounded by the Arctic Circle. The land there is the great tundra region of the world. Tundra is a Russian word which means "marshy plain". For nine months of the year, the tundra is a great white plain. The ground is frozen hard. But in summer the tundra is covered with flowers, thick grass, and moss.

A surprisingly large number of animals live in the tundra area. Reindeer are used for transportation, food, and clothing by the northern peoples. Other animals are polar bears, musk oxen, Arctic hares, and lemmings. In the summer, wolves and foxes migrate to the tundra.

Some of the tundra region is very much like the deserts of the world.

Unit Five Science and Technology in the Middle Ages and Renaissance

To be sure, tundra temperatures are low while desert temperatures are usually high. But in both regions there is little rainfall, not much plant life, and few people.

(1) What topic is treated in this passage?

　　A) The North Pole.

　　B) The animals living in the tundra area.

　　C) The tundra region and the deserts of the world.

　　D) The tundra area within the Arctic Circle.

(2) According to the passage, the great tundra region of the world is _____.

　　A) always a great white plain　　B) a marshy plain of Soviet Russia

　　C) bounded by the Arctic Circle　D) a densely populated area

(3) The animals used for food by the native people are _____.

　　A) bears and musk oxen　　　B) wolves and foxes

　　C) reindeer　　　　　　　　D) Arctic hares and lemmings

(4) The tundra region differs from the deserts of the world in that _____.

　　A) there is much plant life at the tundra while there is not any in the deserts

　　B) there are variety of animals at the tundra while there is no animal in the deserts

　　C) tundra temperatures are low while desert temperatures are usually high

　　D) tundra temperatures are high while desert temperatures are usually low

(5) In Paragraph 3, the author develops his main idea by _____.

　　A) explaining the relationship between the tundra and the deserts

　　B) listing the things the tundra and the deserts share in common

　　C) comparing the tundra with the deserts

　　D) listing the things in which the tundra and the deserts differ from each other

分析：本短文共三段，按重要性顺序排列。第一段为主题段，谈论 Tundra area 的概貌；第二段论述 Tundra area 的动物；第三段比较 Tundra 的部分地区与 deserts 的异同点。

　　在组织段落方面，作者以不同的逻辑方式组织每一段。第一段采用简单排列的方式。全段无主题句，分四个层次。第一层次为第 1 句，阐述 Tundra 的位置；第二层次为第 2 句，谈论 Tundra 的入口；第三层次为第 3 和第 4 句，概述 Tundra 的地貌特征；第四层次是第 5 至 7 句，谈论 Tundra 的气候。这四层次同时指向主题，没有明显主次之分。第二段采用重要性排序方式。第 1 句为主题句。第 2 句重点提及那里最重要的一种动物 reindeer；第 3 句从点到面，泛泛提了一下那里的其他动物。第三段采用比较对照方式。第 1 句主题句；第 2 句通过对照点出 tundra 和 deserts 的相异之处；第 3 句由比较点出两者的相似之处。根据以上分析，题 1 答案为 D。

根据题 2 的题干，该题涉及的有关细节在第一段有关地貌的那一层次，答案可能是 B 或 C，又由该段第一、二层次，进一步确定答案为 C。

根据题 3 的题干，该题有关细节在第二段，答案为 C。

根据题 4 的题干，该题涉及的细节在第三段，不难进一步确定该细节的确切位置，从而得知答案为 C。

根据结构分析，题 5 答案为 C。

在阅读过程中，我们也发现有的文章中事实与细节并不是很明显，我们在提取信息时需要一个再加工的过程，如采用去伪存真的判断、计算等方式。试分析以下例文。

As world population becomes denser, we will feel greater pressure from the expanding number of people. Some experts argue that we are approaching the limit of the number of people the earth can support adequately, and they feel we should turn to compulsory birth control. Other authorities feel that if birth control is imposed on the population, the future of mankind would be seriously jeopardized. They think that very intelligent people would be more likely to have fewer children and this would bring about a lowering of the general level of intelligence in the population as a whole. However some critics see a fallacy in this argument. In addition to genetics, they say, intelligence depends on an adequate diet, a good home environment, parental attention, and education—all of which are increasing in the world as the general education becomes more affluent.

Unit Five   Science and Technology in the Middle Ages and Renaissance

(1) Some people think birth control is _____.
   A) not a workable idea   B) too strict
   C) only for religious people   D) dangerous

(2) Critics say intelligence is based on _____.
   A) a good home environment   B) adequate diet
   C) genetics   D) all of the above

(3) Which of the following is not true?
   A) The whole world is faced with the problem of the increasing population.
   B) Some genetics argue that birth control can affect the level of intelligence.
   C) Genetics is one of the main factors to intelligence.
   D) Food, family attention and education have nothing to do with intelligence.

分析：

　　从短文中可知，题 1 根据短文的相关部分就能找到答案。根据文中第 3 句 "Other authorities feel that if birth control is imposed on the population, the future of mankind would be seriously jeopardized." 和上下文的内容，可以推断出题 1 答案为 A。

　　题 2 属于要求确认短文叙述的事实细节或判断题目陈述的事实或观点的真伪这类题。根据短文中与本题有关部分 "However some critics see a fallacy in this argument..."，A 项、B 项和 C 项均有提及，因此答案为 D。

　　题 3 是一道是非题，主要考核我们辨别是非的能力。这类题要求我们深入理解全文，抓住中心思想和关键词。A 项、B 项和 C 项的内容在短文中都可以找到，属于正确的内容；而 D 项的内容正好与短文最后一段所陈述的内容相反，属于不正确的内容，因此答案为 D。

　　阅读文章的理解分为表层理解和内涵理解。表层理解指要读懂文章的基本内容，包括主要事实和各个细节；而内涵理解指要读懂文章（或作者）通过基本内容表达的思想倾向或主题思想。对文章主要事实和细节的理解属表层理解。阅读理解中针对事实与细节的题目主要考核对组成段落的主体部分的理解程度，检验对中心思想的理解深度及对段落结构组织形式的分析能力。这些事实与细节题可以通过段落层次的规律、组织文章思想的逻辑方式和与中心思想的关系等技巧在文章中找到答案。

Quiz

I. Read each of the following paragraphs and write down the supporting details.

1

Flextime, or flexibility of working hours, has become popular in recent years. Many companies have found that flextime has several advantages. The most obvious advantage is less absenteeism. When employees can choose working hours that meet their needs, they are less likely to take time off. Another advantage is more efficient use of the physical plant. The additional hours that a company is "open for business" could mean higher productivity and greater profits. Finally, giving employees a choice of working hours permits them more control over their work environment, leading to increased job satisfaction and less turnover.

Topic: Flextime

Main Idea: Flextime offers several advantages to companies.

Supporting Details:

(1) _____.
(2) _____.
(3) _____.

2

Changing a career in midlife can be pretty sorry, yet many people are doing just that, and for most it's a good idea. At age 20, who can know what's the best thing to do for the next fifty years? Since people change as they age, a profession that seemed promising at first can become drudgery ten years later. One option for some who want to change careers is to make a business out of a hobby. One man who likes to work with his hands, for example, started a new-

Unit Five   Science and Technology in the Middle Ages and Renaissance

successful handyman service. Others make a midlife change by going to a college or a trade school to begin a new career. They often benefit greatly, as older students bring to their studies experience, judgment and motivation that younger students just don't have.

Topic: Career Change in Mid-life

Main Idea: It is a good idea for many people to change their careers in their mid-life.

Supporting Details:

(1) _____.

(2) _____.

(3) _____.

II. Read each of the following paragraphs and select the best answer.

It was the use of the telescope, of course, that opened the modern age of astronomy and made possible the growth of all our current theories. Johannes Kepler and Tycho Brahe tried to answer some questions about the solar system, but it was Galileo who made the first use of the telescope to observe the heavens close up. Born in 1564, Galileo added greatly to our knowledge of the stars before he died in 1642. By means of his telescope he discovered moons in orbit around Jupiter. Although he saw only four, scientists after him discovered twelve more moons. Galileo also observed that the planet Venus did not always appear the same size. It was his wise use of the telescope that helped him understand this important fact: that the sun and not the earth is the center of the planets.

(1) Circle the topic sentence of the paragraph.

　　A) Sentence 1　　　　　　B) Sentence 7

C) Sentence 5   D) Sentence 6

(2) Which of the following facts are important in this paragraph?

A) Kepler and Brahe answered questions about the solar system.

B) Galileo was born in 1564 and died in 1642.

C) Galileo discovered moons around Jupiter.

D) Galileo observed the planet Venus.

E) Galileo came to understand that the earth is not the center of the stars' system.

(3) Galileo's major contribution to our knowledge of the planet was _____.

A) his invention of the telescope

B) his study on Brahe and Kepler

C) his understanding that planets revolve around the sun

D) his observations about Jupiter and Venus

Part B Reading

Medieval Science and Technology

Science in the Middle Ages comprised the study of nature, including practical disciplines, the mathematics and natural philosophy in medieval Europe. Following the fall of the Western Roman Empire and the decline in knowledge of Greek, Christian Western Europe was cut off from an important source of ancient learning. Although a range of Christian clerics and scholars maintained the spirit of rational inquiry, Western Europe[1] would see during the Early Middle Ages (476–1000 A.D.) a period of intellectual stagnation. During the High Middle Ages (1000–1300 A.D.), however, the West had begun to reorganize itself and was on its way to taking again the lead in scientific discovery. Renaissance science spawned the Scientific Revolution; science and technology began a cycle of mutual advancement.

Medieval science

An intellectual revitalization of Europe started with the birth of medieval universities in the 12th century. The contact with the Islamic world in Spain and Sicily, and during the Reconquista[2] and the Crusades[3], allowed Europeans access to scientific Greek and Arabic texts, including the works of Aristotle, Ptolemy, and others. European scholars had access to the translation programs of Raymond of Toledo, who sponsored the 12th century Toledo School of Translators[4] from Arabic to Latin. Later translators like Michael Scotus would learn Arabic in order to study these texts directly. The European universities aided materially in the translation and propagation of these texts and started a new infrastructure which was needed for scientific communities. In fact, European university put many works about the natural world and the study of nature at the center of its curriculum, with the result that the "medieval university laid far greater emphasis on science than does its modern counterpart and descendant".

As well as this, Europeans began to venture further and further east as a result of the *Pax Mongolica*[5]. This led to the increased influence of Indian and even Chinese science on the European tradition. Technological advances were also made, such as the early flight of Eilmer of Malmesbury[6] (who had studied Mathematics in the 11th-century England), and the metallurgical achievements of the Cistercian[7] blast furnace at Laskill.

At the beginning of the 13th century there were reasonably accurate Latin translations of the main works of almost all the intellectually crucial ancient authors, allowing a sound transfer of scientific ideas via both the universities and the monasteries. By then, the natural philosophy contained in these texts began to be extended by notable scholastics such as Robert Grosseteste[8], Roger Bacon[9], Albertus Magnus[10] and Duns Scotus[11]. Precursors of the modern scientific method, influenced by earlier contributions of the Islamic world, can be seen already in Grosseteste's emphasis on mathematics as a way to understand nature, and in the empirical approach admired by Bacon, particularly in his *Opus Majus*[12]. Pierre Duhem's[13] provocative thesis of the Catholic Church's Condemnation of 1277[14] led to the study of medieval science as a serious discipline, "but no one in the field any longer endorses his view that modern science started in 1277".

The first half of the 14th century saw much important scientific work being done, largely within the framework of scholastic commentaries on Aristotle's scientific writings. William of Ockham[15] introduced the principle of parsimony: natural philosophers should not postulate unnecessary entities, so that motion is not a distinct thing but is only the moving object and an intermediary "sensible species" is not needed to transmit an image of an object to the eye. Scholars such as Jean Buridan[16] and Nicole Oresme[17] started to reinterpret elements of Aristotle's mechanics. In particular, Buridan developed the theory that impetus was the cause of the motion of projectiles, which was a first step towards the modern concept of inertia.

Unit Five   Science and Technology in the Middle Ages and Renaissance

The Oxford Calculators began to mathematically analyze the kinematics of motion, making this analysis without considering the causes of motion.

In 1348, the Black Death [18] and other disasters sealed a sudden end to the previous period of massive philosophic and scientific development. Yet, the rediscovery of ancient texts was improved after the Fall of Constantinople [19] in 1453, when many Byzantine scholars had to seek refuge in the West. Meanwhile, the introduction of printing was to have great effect on European society. The facilitated dissemination of the printed word democratized learning and allowed a faster propagation of new ideas. New ideas also helped to influence the development of European science at this point: not least the introduction of Algebra [20]. These developments paved the way for the Scientific Revolution [21], which may also be understood as a resumption of the process of scientific change, halted at the start of the Black Death.

Medieval technology

After the Renaissance of the 12th century [22], medieval Europe saw a radical change in the rate of new inventions, innovations in the ways of managing traditional means of production, and economic growth. The period saw major technological advances, including the adoption of gunpowder, the invention of vertical windmills, spectacles, mechanical clocks, and greatly improved water mills, building techniques (Gothic style, medieval castle), agriculture in general (three-field crop rotation).

The development of water mills from its ancient origins was impressive, and extended from agriculture to sawmill both for timber and stone. By the time of the Domesday Book [23], most large villages had turnable mills, around 6500 in England alone. Water-power was also widely used in mining for raising ore from shafts, crushing ore, and even powering bellows.

European technical advancements in the 12th to 14th centuries were either built on long-established techniques in medieval Europe, originating from Roman and Byzantine antecedents, or adapted from cross-cultural exchanges through trading networks with the Islamic world, China, and India. Often, the revolutionary aspect lay not in the act of invention itself, but in its technological refinement and application to political and economic power. Though gunpowder had long been known to the Chinese, it was the Europeans who developed and perfected its military potential, precipitating European expansion and eventual imperialism in the Modern Era.

Also significant in this respect were advances in maritime technology. Advances in shipbuilding included the multi-masted ships with lateen sails, the sternpost-mounted rudder and the skeleton-first hull construction. Along with new navigational techniques such as the dry compass [24], the Jacob's staff and the astrolabe, these allowed economic and military control of all seas adjacent to Europe and enabled the global navigational achievements of the dawning Age of Exploration [25].

At the turn to the Renaissance, Gutenberg's [26] invention of mechanical printing made possible a dissemination of knowledge to a wider population, that would not only lead to a gradually more egalitarian society, but one more able to dominate other cultures, drawing from a vast reserve of knowledge and experience. The technical drawings of late medieval artist-engineers Guido da Vigevano [27] and Villard de Honnecourt [28] can be viewed as forerunners of later Renaissance works such as Taccola [29] or da Vinci [30].

New Words and Expressions

cleric 教会圣职人员
stagnation 停滞；淤塞
spawn 造成，酿成；大量产生
infrastructure 基础；基础结构
metallurgical 冶金的；冶金术的
monastery 修道院；庙宇，寺院

precursor 先驱；前辈
provocative 挑衅的；引起争论的
endorse 赞同；认可
commentary 注解（本）；评注
parsimony 吝啬；（在用词方面的）过度节省

Unit Five Science and Technology in the Middle Ages and Renaissance

intermediary 中间的；媒介的	antecedent 祖先
impetus 动力，原动力，推动力	precipitate 促使，加速
projectile 抛射体；发射体	mast 在……上装桅杆
kinematics [物] 运动学，动力学	lateen（地中海上帆船用的）三角帆
dissemination 传播（思想、理论等）；散布	sternpost 船尾骨
halt 停住，停止	hull 船壳，船体
sawmill 锯木厂	egalitarian 平等主义者

Notes

1. Western Europe 指的是含罗马天主教和拉丁语言在内的欧洲文化范围。

2. Reconquista（Reconquest，或称作 the Recapturing）收复失地运动。公元 718—1492 年间，位于西欧伊比利亚半岛北部的基督教各国逐渐战胜南部穆斯林摩尔人政权。史学家以公元 718 年，倭马亚阿拉伯政权征服西哥特王国，以及阿斯图里亚斯王国建国作为收复失地运动的开端，以 1492 年格拉纳达的陷落为终。这个事件的西班牙语和葡萄牙语名称 "Reconquista" 一词有 "重新征服" 的意思。

3. Crusades（1096—1291）十字军东征是西方基督教徒组织反对穆斯林国家的持续近 200 年的宗教性战争。参加这场战争的士兵佩有十字标志，因此称为十字军。十字军东征对西方基督教世界造成了深远的社会、经济和政治影响，其中有些痕迹至今尚存。

4. Toledo School of Translators 托莱多翻译院。通常指 12-13 世纪期间在托莱多城市的一群学者聚集在一起，共同翻译古阿拉伯、古希腊和古希伯来语的哲学和科学著作。

5. The *Pax Mongolica* 拉丁词组，意为 "蒙古和平"（Mongol Peace）。该词由西方学者创造，用来描述 13-14 世纪欧亚内地的大部分随着蒙古的大范围征服所带来的相对和平时期。蒙古征服对东西经济、文化交流有促进之功，也间接启发了之后欧洲人的地理大发现乃至文艺复兴的开端。

6. Eilmer of Malmesbury 马姆斯伯里修道院的埃默，11 世纪英国教士，以首次尝试戴翼滑翔飞行而出名。他自己曾做了一对翅膀，然后从修道院塔顶跳出，飞了约 200 米。

7. Cistercians（简称 OCist）熙笃会（或译西多会）是一个天主教隐修会。熙笃会主张生活严肃，重个人清贫，终身吃素，每日凌晨即起身祈祷。他们在黑色法衣里面穿一件白色会服，所以有时也被称作白衣修士。

8. Robert Grosseteste（约 1175—1253）罗伯特·格罗斯泰斯特，英国政治家、经院哲学家、神学家和伦敦大主教。

9. Roger Bacon（1214—1292）罗吉尔·培根，英国唯物主义思想家和伟大的科学家，著有《大著作》等。

10. Albertus Magnus 艾尔伯图斯·麦格努斯，德国理论家、主教和科学家。博学多才，被誉为"百科学博士"。

11. Duns Scotus（约 1265—1308）董思高，苏格兰中世纪时期的经院哲学家、神学家和唯名论者。他提出了物质具有思维能力的推测，其论据是天主是万能的，故而可以让物质具备思维能力。著有《巴黎论著》《牛津论著》《问题论丛》等。

12. The *Opus Majus*（Latin for "Greater Work"）《大著作》，由罗吉尔·培根所著。

13. Pierre Duhem（1861—1916）杜恒，法国物理学家、数学家和科学哲学家。

14. Condemnation of 1277（1277 年）制裁书。巴黎主教丹皮尔（Etienne Tempier）于 1270 年和 1277 年颁布了两次制裁书（The Condemnations of 1270 and 1277）。

15. William of Ockham（约 1285—1349）奥卡姆的威廉，出生于英格兰的萨里郡奥卡姆（Ockham）。14 世纪逻辑学家和圣方济各会修士。他能言善辩，被人称为"驳不倒的博士"。主要著作有《箴言书注》《逻辑大全》《辩论集 7 篇》等。

16. Jean Buridan（1292—1363）让·布里丹，法国哲学家、经院哲学博士和欧洲宗教怀疑主义倡导者。1340 年，在西方再造了冲力说理论。

17. Nicole Oresme（约 1320—1382）尼克尔·奥里斯姆，中古晚期最

Unit Five   Science and Technology in the Middle Ages and Renaissance

知名、最具影响力哲学家之一。本身为经济学家、数学家、物理学家、天文学家、哲学家、心理学家、音乐学家和神学家等，也是近代科学主要奠基者之一。

18. Black Death 黑死病是人类历史上最严重的瘟疫之一，刚开始被当时的作家称作"Great Mortality"，瘟疫爆发之后，又有了"黑死病"之名。一般认为这个名称是取自其中一个显著的症状，称作"acral necrosis"，患者的皮肤会因为皮下出血而变黑。约在1340年散布到整个欧洲，而"黑死病"之名是当时欧洲的称呼。据估计，瘟疫爆发期间中世纪欧洲约有占人口总数30%的人死于黑死病。黑死病对欧洲人口造成了严重影响，改变了欧洲的社会结构，动摇了当时支配欧洲的罗马天主教会的地位，并因此使得一些少数族群受到迫害。

19. Fall of Constantinople 君士坦丁堡的陷落。它是奥斯曼帝国于苏丹穆罕默德二世领导之下对拜占庭帝国首都君士坦丁堡的一次征服，发生于1453年5月29日星期二。它不但标志着东罗马帝国最后的毁灭以及最后的拜占庭皇帝君士坦丁十一世的死亡，也代表奥斯曼对地中海东部及巴尔干半岛的统治在战略上一次决定性的成功。1922年帝国崩溃前，君士坦丁堡仍然为帝国首都，1930年土耳其共和国官方将其改名为伊斯坦布尔。

20. Algebra 代数学是研究数、数量、关系与结构的数学分支。代数学的研究对象不仅是数字，而是各种抽象化的结构。常见的代数结构类型有群、环、域、模、线性空间等。

21. Scientific Revolution 科学革命大约于1543年开始。那一年尼古拉斯·哥白尼出版了著作《天体运行论》(*De revolutionibus orbium coelestium*)，安德烈·维赛留斯出版了《人体构造》(*De humani corporis fabrica*)。尽管科学革命的具体时间仍有争议，但公认的是，在16-17世纪之间，物理学、天文学、生物学、医学以及化学的思想都经历了根本性的变化，由中世纪的观点转变为现代科学的基础。

22. Renaissance of the 12th century 12世纪的文艺复兴。欧洲中世纪早期和中期，社会、政治、经济发生剧变，哲学和科学的源头重新被找回。这一段时期为15世纪意大利文学和艺术的变化以及17世纪的科学革命铺路。

23. Domesday Book 末日审判书是1086年完成的大规模调查英格兰的

记录，由征服者威廉实施，类似于现在政府的人口普查。征服者威廉需要得到他刚刚征服的国家的信息，以便管理英格兰。调查的主要目的是找出谁拥有什么并让他们交税。书名 Domesday（Doomsday 的中古英语拼法，意为"世界末日"）从 12 世纪开始使用，强调了这本书的最终性和权威性。而根据其中的调查结果，英格兰约有 150 万人口，其中 90% 以上是农民。

24. dry compass 旱罗盘。旱航海罗盘约在 1300 年在欧洲发明。

25. Age of Exploration 欧洲历史的地理大发现，又名探索时代或大航海时代。指从 15 世纪到 17 世纪时期，欧洲的船队出现在世界各处的海洋上，寻找着新的贸易路线和贸易伙伴，以发展欧洲新生的资本主义。在这些远洋探索中，欧洲人发现了许多当时在欧洲不为人知的国家与地区。与此同时，欧洲涌现出了许多著名的航海家，如哥伦布、麦哲伦等。伴随着新航路的开辟，东西方之间的文化、贸易交流开始大量增加，殖民主义与自由贸易主义也开始出现。欧洲这个时期的快速发展奠定了其繁荣程度超过亚洲的基础。新航路的发现，对世界各大洲在数百年后的发展也产生了久远的影响。

26. Johannes Gutenberg（1398—1468）约翰尼斯·谷登堡，第一位发明活字印刷术的欧洲人。1455 年他发明了铅活字凸版机械印刷机。他的发明引发了一次媒介革命。

27. Guido da Vigevano 维杰瓦诺，意大利内科医生和发明家，以描绘大量的技术条目和精巧的设备的素描本而著名。

28. Villard de Honnecourt 维拉尔德·奥内库尔，来自法国南部庇卡底（Picardy）的 13 世纪的艺术家，因其现存的含有 33 张描写降落伞的文件而出名。

29. Taccola 塔科拉，早期文艺复兴时期意大利管理者、艺术家和工程师，因其艺术专著而出名。

30. Da Vinci（1452—1519）达·芬奇，意大利文艺复兴三杰之一，也是整个欧洲文艺复兴时期最完美的代表之一。

Unit Five   Science and Technology in the Middle Ages and Renaissance

📝 Exercises

I. Find out the English equivalents to the following Chinese terms from the passage.

1. 理性求索的精神
2. 知识停滞期
3. 知识复元
4. 中世纪大学的诞生
5. 科学群体
6. 跨文化交流
7. 技术上的精心改进
8. 航海技术
9. 旱罗盘
10. 罗盘支杆

II. Answer the following questions by using the sentences in the passage or in your own words.

1. Can you explain why "medieval university laid far greater emphasis on science than did its modern counterpart and descendant"?
2. The first half of the 14th century saw much important scientific work being done. What do you know about it?
3. What contributed to the European technical advancements in the 12th to 14th centuries?

Part C Extended Reading

Johannes Gutenberg and His Invention of Movable Type Printing

"What the world is today, good and bad, it owes to Gutenberg. Everything can be traced to this source, but we are bound to bring him homage, ...for the bad that his colossal invention has brought about is overshadowed a thousand times by the good with which mankind has been favored."

—— American writer Mark Twain (1835–1910)

The invention of mechanical movable type printing by the German goldsmith Johannes Gutenberg (1398–1468) started the Printing Revolution and is widely regarded as the most important event of the modern period. It played a key role in the development of the Renaissance, Reformation[1], the Age of Enlightenment and the Scientific Revolution and laid the material basis for the modern knowledge-based economy and the spread of learning to the masses.

Gutenberg was the first European to use movable type printing, in around 1439. But his technique of making movable type remains unclear. In the following decades, punches and copper matrices became standardized in the rapidly disseminating printing presses across Europe. In the standard process of making type, a hard metal punch (made by punchcutting, with the letter carved back to front) is hammered into a softer copper bar, creating a *matrix*. This is then placed into a hand-held mould and a piece of type, or "sort", is cast by filling the mould with molten type-metal; this cools almost at once, and the resulting piece of type can be removed from the mould. The matrix can be reused to create hundreds, or thousands, of identical sorts so that the same character appearing anywhere within the book will appear very uniform, giving rise,

Unit Five   Science and Technology in the Middle Ages and Renaissance

over time, to the development of distinct styles of typefaces or fonts. After casting, the sorts are arranged into type-cases, and used to make up pages which are inked and printed, a procedure which can be repeated hundreds, or thousands, of times. The sorts can be reused in any combination, earning the process the name of "movable type".

Gutenberg made many contributions to printing, for instance, the invention of a process for mass-producing movable type; the use of oil-based ink; and the use of a wooden printing press similar to the agricultural screw presses of the period. His truly epochal invention was the combination of these elements into a practical system which allowed the mass production of printed books and was economically viable for printers and readers alike. Gutenberg's method for making type is traditionally considered to have included a type metal alloy and a hand mould for casting type.

The use of movable type was a marked improvement on the handwritten manuscript, which was the existing method of book production in Europe, and upon woodblock printing, and revolutionized European book-making. Gutenberg's printing technology spread rapidly throughout Europe and later the world.

His major work, the *Gutenberg Bible*[2](also known as the *42-line Bible*), has been acclaimed for its high aesthetic and technical quality.

In modern times, Gutenberg remains a towering figure in the popular image. In 1999, the A&E Network ranked Gutenberg the No. 1 most influential person of the second millennium on their "Biographies of the Millennium" countdown. In 1997, *Time-Life* magazine picked Gutenberg's invention as the most important of the second millennium.

New Words and Expressions

homage 尊敬；敬意；崇敬
colossal 异常的，非常的
punch （凸模）冲头，阳模
matrix （复数为 matrices）[印] 字模；（铸铅版用的）纸型
type （总称；也可作可数名词，指一个）[印] 铅字，活字
mould 模型；铸模
cast 铸（造）
molten 熔融的，熔化的

uniform（与其他）一样的，相同的
typeface 铅字面；铅字印出的字样
font 字型，字体
epochal 新纪元的，划时代的，有重大意义的
viable 可行的
acclaim 向……欢呼，为……喝彩；以欢呼声承认
towering 屹立的，高耸的；巨大的
countdown 以倒数方式进行的时间计算

Notes

1. the Reformation 指（16世纪欧洲的）基督教改革运动。
2. *Gutenberg Bible* 谷登堡所著《谷登堡圣经》，又称为《42行圣经》。

Exercises

I. Find out the English equivalents to the following Chinese terms from the passage.

1. 活字
2. 铜质型片
3. 活字金属/字铅
4. （汉）字
5. 铅字盘
6. 浇铸铅字
7. 手写稿
8. 木版印刷
9. 高大的人物
10. 大众印象

II. Translate the following sentences into Chinese.

1. The invention of mechanical movable type printing by the German goldsmith Johannes Gutenberg (1398–1468) started the Printing Revolution and is widely regarded as the most important event of the modern period. It played a key role in the development of the Renaissance, Reformation, the Age of Enlightenment and the Scientific Revolution and laid the material basis for the modern knowledge-based economy and the spread of learning to the masses.

Unit Five   Science and Technology in the Middle Ages and Renaissance

2. His truly epochal invention was the combination of these elements into a practical system which allowed the mass production of printed books and was economically viable for printers and readers alike.
3. His major work, the *Gutenberg Bible* (also known as the *42-line Bible*), has been acclaimed for its high aesthetic and technical quality.

Unit Six
Impact of Science in Europe

Part A Lecture

识别逻辑衔接词

语篇是由句子、句群组成的，但不是句子的任意堆砌。语篇是一个语义单位，构成语篇的句子必须是相关的；它们不仅在意义上相关，同属一个语义场，而且在结构上也相关，通过一定的联结手段合乎逻辑地连在一起。语篇成为一个意义连贯的有机整体。达到语篇语义衔接的手段很多，使用逻辑衔接词（logical connectors）是其中之一。识别和依靠句子之间的逻辑衔接词的能力是一项非常重要的阅读技巧，有助于我们弄清文章层次，超越句子水平理解文章意义并获取信息。本单元讲座将详细阐述逻辑衔接词在句子之间、段落内和段落之间的作用。

1 识别逻辑衔接词的阅读技巧

逻辑衔接词是表示各种逻辑意义的词、短语或分句，它们主要用来表示两个或更多的句子之间的某种逻辑关系。表示逻辑意义的词主要有连词和副词，如 and、but、for、then、yet、so、therefore、however 等；短语主要为介词短语，如 in addition、as a result、on one hand、on the other hand、on the contrary、in other words、for example 等；分句包括非限定分句和限定分句，如 considering all that、to conclude、that is to say、what is more、what is more important 等。阅读下面短文，指出表示逻辑意义的词和短语。

The most parts of the United States, wilderness areas are steadily decreasing in size because of the pressure of a growing population. Forests and deserts alike have turned into suburban housing developments. In some of the north eastern states, however, the reverse is happening. As

Unit Six Impact of Science in Europe

unprofitable farms are abandoned, their pasturelands revert to forest. For example, fifty years ago, only 30 percent of the land in Vermont was forest; today that figure is closer to 70 percent. As a result, many animals that had declined in number or had disappeared completely from the area are now returning. Black bears, red wolves, beavers, and even coyotes from the western states can now be found in large numbers throughout northern New England.

分析：以上短文中用来表示句子之间逻辑关系的词和短语有：句 1 的 because of，句 3 的 however，句 5 的 For example，句 6 的 As a result，句 7 的 and。

从上例可知，逻辑衔接词是用来表示句子之间逻辑关系的，这些逻辑衔接词本身不能直接影响前一句或后一句的结构，但它们的具体意义表明必须有其他句子存在。如下面例句：

A. **In addition**, he is considerate towards others in class.

B. **However**, he had tried his best.

分析：句 A 的逻辑衔接词 in addition 表达了一种递进连接关系，说明句子前面应还有其他句子存在，否则这个句子就不能成立。同样，句 B 的逻辑衔接词 however 是转折连接，它前面也应有其他句子与其连接。因此，这两个例句可改为：

A. He is intelligent and hardworking. **In addition**, he is considerate towards others in class.（他聪明且勤奋。此外，他对班上同学也很体贴。）

B. He failed. **However**, he had tried his best.（他失败了，但是他尽了最大的努力。）

从以上例子可以看出，逻辑衔接词所表达的是语义上的关系，而不是语法关系。由此也可看出，逻辑衔接词的功能是表示两个或更多的句子之间的某种逻辑关系，并指出句子是在什么意义上相互联系起来的；同时也用于表明语篇中的各个组成部分之间语义上的关联。

那么，逻辑衔接词表达了什么语义关系呢？从意义上看，逻辑衔接词可以表示多种不同的语义关系，这些不同的语义关系主要分为以下六大类：

1.1 表示递进关系

表递进关系的逻辑衔接词连接前后两个相同思维的句子，衔接词后面的内容往往是对前面内容的增补。如：

Finally, a knowledge of statistics is required by every type of scientists for the analysis of data. **Moreover**, even an elementary knowledge of this branch of mathematics is sufficient to enable the journalist to avoid misleading his readers, or the ordinary citizen to detect the attempts which are constantly made to deceive him.

分析：例中 moreover 的前一部分说明"统计学的知识是各行各业的科学家分析数据时必须具备的"，moreover 后面的内容"对于新闻记者来说，只要稍有一点统计学知识，就可以避免因为报道失实而将读者引入歧途。而对于普通老百姓来说，则可以利用这方面的知识，来识破往往使之上当的不良企图。"是对前面内容的增补。

表示递进关系常用的逻辑衔接词有：and、and also、furthermore、moreover、besides、in addition、what is more 等。

1.2 表示转折关系

表转折关系的逻辑衔接词主要连接前后两个互为相反的思维方向。如：

There are two kinds of memory: short-term and long-term. Information in long-term memory can be recalled at a later time when it is needed. It may be kept for days, weeks or months. **In contrast**, information in short-term memory is kept for a few seconds, usually by repeating the information over and over.

分析：本段指出"有长期记忆和短期记忆两种记忆方式"。而后通过 In contrast 比较了两者的不同，即"长期记忆中的信息在以后需要的时候可以回忆起来，可以为期几天、几个星期乃至几个月。而短期记忆中的信息只能维持数秒，需要再三重复"。

表示转折关系常用的逻辑衔接词有：but、yet、however、although、in spite of、despite、otherwise、instead、rather、on the contrary、by contrast、on the one hand...on the other hand、as a matter of fact 等。

1.3 表示因果关系

表因果关系的逻辑衔接词主要说明什么样的原因导致什么样的结果，或引起某一结果或现象的原因是什么。如：

To most people the term technology conjures up images of smoky steel mills or noisy machines. Perhaps the classic representation of technology is still the assembly line created by Henry Ford half a century ago and made into a social symbol by Charlie Chaplin in *Modern Times*. This symbol, however, has always been inadequate and misleading, **for** technology has always been more than factories and machines.

分析：例句中 for 后面的内容告诉我们，技术这一象征不完善且易引起误解的原因——技术从来就不仅仅是工厂和机器。

表示因果关系常用的逻辑衔接词有：for、as、since、because、for this reason、because of this、owing to、due to、hence、so、thus、therefore、as a result、accordingly、consequently、in consequence、in such an event、on account of this、that being the case 等。

表示因果关系的逻辑衔接词可以分为以下两种：

▶ 1.3.1 表示原因的逻辑衔接词

Most nutrition experts today would recommend a balanced diet containing elements of all foods. It's largely **because** we need sufficient vitamins. Vitamins were first called "accessory food factors" **since** it was discovered that most foods contain besides carbohydrates, fats, minerals and water, some other substances necessary for health.

分析：句中 because 和 since 表示原因，意思是"现在大多数营养学家都推荐含有多种食物元素的平衡营养。这在很大程度上由于我们需要足够的维生素。维生素起先称之为'附加食物因子'，因为人们发现大多数食物除含有碳水化合物、脂肪、矿物质、水以外，还有人体必须的其他物质"。

1.3.2 表示结果的逻辑衔接词

It has been observed that high salt consumption is linked to hypertension. **Consequently**, most clinicians advise patients to lower sodium intake to mitigate this risk.

分析：第 1 句告知高盐摄入量与高血压相关，结果是（Consequently）为减轻高血压风险，大多数临床医生提醒患者降低钠的摄入量。

1·4 表示解释关系

表示解释关系的逻辑衔接词连接前后两个相同思维的句子，衔接词后面的内容往往是对前面内容的解释和举例。如：

Even a detailed account of the history of science cannot be complete, for scientific activity is not isolated but takes place within a larger matrix that also includes, **for example**, political and social events, developments in the arts, philosophy, and religion, and forces within the life of the individual scientist.

分析：例中 for example 的前一部分阐述"科学活动不是孤立的，而是发生在更大的矩阵里"，for example 后面的内容具体解释这个矩阵包括的内容，给出诸如"政治和社会事件，艺术、哲学和宗教的发展，科学家个人生活中的各种势力"的实例。

表示解释关系常用的逻辑衔接词有：that is、that is to say、namely、in other words、similarly、likewise、for example、for instance 等。如：

Unit Six　Impact of Science in Europe

All of these areas of study might be called pure sciences, in contrast to the applied, or engineering, sciences, **i.e.**, technology, which is concerned with the practical application of the results of scientific activity.

分析：句中 i.e.（that is）起解释作用，解释"应用科学"或称为"工程科学"，也就是"技术"。

1·5　表示列举关系

表列举关系的逻辑衔接词主要列举动作发生的顺序或信息间的顺序。如：

Technological innovation consists of three stages, linked together into a self-reinforcing cycle. **First**, there is the creative, feasible idea. **Second**, its practical application. **Third**, its diffusion through society.

分析：句中的 first、second 和 third 用来说明动作发生的先后顺序。

表示列举关系常用的逻辑衔接词有：first、second、third...、for one thing...for another、in the first place、to begin with、initially、and so on、then、next、before、after、last、afterward、previously、finally、at once、soon、on another occasion、an hour later、at this point 等。

1·6　表示总结关系

表示总结关系逻辑衔接词主要说明结论或总结的所在之处。如：

Given a certain power of engine, and consequently a certain fuel consumption, there is a practical wanted for fuel, radio navigational instruments, passengers seats, or freight room, and, of course, the passengers or freight themselves. **So** the structure of the aircraft has to be as small and light as safety and efficiency will allow.

分析：例句中的 so 表明其后面的内容是结论——在安全和效率允许的情况

下，飞机的结构必须尽可能地做到小且轻。

表示总结关系常用的逻辑衔接词有：(all) in all、in conclusion、to conclude、in short、in brief、on the whole、to sum up、therefore、in a word、so、as a result、finally、accordingly、thus、consequently、above all 等。

2 综合练习

语篇中句与句之间存在一定的逻辑关系，表示这些逻辑关系的逻辑衔接词表示递进、转折、因果、解释、列举、总结等各种不同的语义关系。上节我们主要介绍了逻辑衔接词在句间的作用。本节主要了解逻辑衔接词在段落内以及段落与段落之间的作用，练习识别段落中及段落间的逻辑衔接词。

2·1 识别逻辑衔接词的作用

阅读下面段落，指出段中斜体逻辑衔接词的语义关系。

Life in a new country can be confusing. **For example**, one day I wanted to go to the consulate to renew my student visa. **So** my aunt gave me the consulate's address. **But** when I arrived downtown, I got lost. **First**, I went to a bank. A lady there told me to walk three blocks south. **However**, I walked three blocks north instead. **Then** I asked a policeman for help. **As a result**, he drove me right to the door of the consulate. **From my story**, you can see that I had a hard time doing one errand.

分析：本段的逻辑衔接词及其语义关系如下：

① 解释关系，表示举例，紧接 for example 的内容是具体的例证。

② 因果关系，表示结果。

③ 转折关系，表示相反的结果。

④ 列举关系，表示动作发生的第一步。

⑤ 转折关系，表示前后内容不一致。

Unit Six　Impact of Science in Europe

⑥ 列举关系，表示动作发生的第二步。
⑦ 因果关系，告知结果。
⑧ 总结关系，说明结论之所在。

2·2 分析逻辑衔接词的作用

阅读下面段落，试分析逻辑衔接词在段内的作用。

　　The law of conservation of energy states that energy can be neither created nor destroyed. **However**, energy can be converted from one form to another. Electrical energy, **for instance**, can be converted into heat energy—**that is** what happens when you turn on your toaster.

分析：本段第 1 句告诉我们能量守恒定律。第 2 句 However 表示转折，突出其为主题句，表示"但能量是可以相互转换的"，并通过第 3 句的 for instance 举例"电能可以转换成热能"。第 3 句中的 that is 表示解释。

　　Let us look at three categories of diseases which are related to food. **First**, people may get a disease because they lack a nutrient they need. **For example**, if they do not have enough iron, they will get a form of anemia. **Second**, people may get food poisoning. There are four major sources of food poisoning: harmful bacteria, parasites (like worms), harmful chemicals, and naturally poisonous plants **such as** some mushrooms. **Finally**, some people have food allergics. These people are very sensitive to certain foods. **For instance**, some people get sick when they eat eggs because they are allergic to them.

分析：本段阐述与食品有关的三种疾病。段中使用 first、second、finally 逻辑衔接词列举了生病的原因，如缺乏需要的营养、食物中毒、对某些食品过敏；并且使用了 for example、such as、for instance 等逻辑衔接词引出例子。本段意为"让我们看看三种与食品有关的疾病。首先，人们生病因为缺乏他们需要的营养。例如，如果他们缺乏足够的铁，他们就会得贫血。其次，人们会食物中毒。食物中毒有四种原因：有毒细菌、寄生虫、有害化学物质和自然界中的毒草，如毒蘑菇。最后，有些人具有食物过敏症。这些人对某些食品过敏，例如，有些人吃了鸡蛋就会得病，因为他们对鸡蛋过敏"。

从上述例子可以看出，句子与句子的语义连贯主要通过表示各种关系的逻辑联系语达到的。因此，理解逻辑联系语的语义关系能帮助理解文章的发展脉络，理解文章的语篇框架，从而理解文章的主要内容。

Quiz

Read the following paragraphs. As you read, study the way the words in bold are used to show relationships among ideas within the paragraph. And then indicate the idea each word expresses in bold.

(1) The growing number of computer crimes shows that computers can be used to steal. Computer criminals do not use guns. And **even if** they are caught, it is hard to punish them **because** there are no witness and often no evidence.

(2) There are three main ways that vitamins are lost from foods. **First**, some vitamins dissolve in water. When vegetables are cooked in water, the vitamins go into the water. If the water is thrown away, the vitamins are lost. **In addition**, heat, light, and oxygen change some vitamins. If vegetables are cooked, the heat changes some of the vitamins. Then these vitamins are no longer healthful. **For instance**, broiling grapefruit destroys some of its vitamin C. **Finally**, people sometimes throw away the part of a plant which has the most vitamins. **For example**, when flour is processed, the wheat germ, which is very rich in vitamins, is lost.

(3) There are two factors which determine an individual's intelligence. **The first is** the sort of brain he is born with. Human brains differ considerably, some being more capable than others. **But** no matter how good a brain he has to begin with, an individual will have a low order of intelligence unless he has opportunities to learn. **So the second factor is** what happens

Unit Six   Impact of Science in Europe

to the individual—the sort of environment in which he is reared. If an individual is handicapped environmentally, it is likely his brain will fail to develop and he will never attain the level of intelligence of which he is capable.

(4) In the more developed countries, automation will lead to a high degree of efficiency in the production of manufactured goods, and is likely to have far-reaching social effects. **For instance**, workers will need to be more highly trained and more flexible—they will probably have to be capable of changing from one skilled job to another—and they will also have more free time, as they will work fewer hours per day. This in turn will necessitate a considerable expansion and re-orientation of education. **Another** result of automation should speed up the accumulation of surplus capital, which could then be made available for the purpose of assisting the emerging countries to solve some of the problems of underdevelopment.

Part B Reading

New Ideas During the Scientific Revolution

The scientific revolution is an era associated primarily with the 16th and 17th centuries during which new ideas and knowledge in physics, astronomy, biology, medicine and chemistry transformed medieval and ancient views of nature and laid the foundations for modern science.

The scientific revolution was not marked by any single change. The following new ideas contributed to what is called the scientific revolution:

• The replacement of the Earth as center of the universe by Heliocentrism.

• Deprecation of the Aristotelian theory[1] that matter was continuous and made up of the elements Earth[2], Water, Air, and Fire because its classic rival, Atomism, better lent itself to a "mechanical philosophy" of matter.

• The replacement of the Aristotelian idea that heavy bodies, by their nature, moved straight down toward their natural places; that light bodies, by their nature, moved straight up toward their natural place; and that ethereal bodies, by their nature, moved in unchanging circular motions with the idea that all bodies are heavy and move according to the same physical laws.

• Inertia replaced the medieval impetus theory[3], that unnatural motion ("forced" or "violent" rectilinear motion) is caused by continuous action of the original force imparted by a mover into that which is moved.

• The replacement of Galen's[4] treatment of the venous and arterial systems as two separate systems with William Harvey's[5] concept that blood circulated from the arteries to the veins "impelled in a circle, and is in a state of ceaseless motion".

Unit Six   Impact of Science in Europe

Many of the important figures of the scientific revolution, however, shared in the Renaissance respect for ancient learning and cited ancient pedigrees for their innovations. Nicolaus Copernicus [6] (1473–1543), Kepler [7] (1571–1630), Newton [8] (1642–1727) and Galileo Galilei [9] (1564–1642) all traced different ancient and medieval ancestries for the heliocentric system. In the Axioms Scholium of his *Principia* Newton said its axiomatic three laws of motion [10] were already accepted by mathematicians such as Huygens [11] (1629–1695), Wallace, Wren and others, and also in memos in his draft preparations of the second edition of the *Principia* he attributed its first law of motion and its law of gravity to a range of historical figures. According to Newton himself and other historians of science, his *Principia*'s first law of motion was the same as Aristotle's counterfactual principle of interminable locomotion in a void stated in *Physics* and was also endorsed by ancient Greek atomists and others.

The geocentric model [12] was nearly universally accepted until 1543 when Nicolaus Copernicus published his book entitled *De revolutionibus orbium coelestium* [13] and was widely accepted into the next century. At around the same time, the findings of Vesalius corrected the previous anatomical teachings of Galen, which were based upon the dissection of animals even though they were supposed to be a guide to the human body.

Andreas Vesalius [14] (1514–1564) was an author of one of the most influential books on human anatomy, *De humani corporis fabrica* [15], also in 1543. French surgeon Ambroise Paré [16] (c.1510–1590) is considered as one of the fathers of surgery; he was leader in surgical techniques and battlefield medicine, especially the treatment of wounds. Partly based on the works by the Italian surgeon and anatomist Matteo Realdo Colombo [17] (c. 1516–1559), the anatomist William Harvey (1578–1657) described the circulatory system. Herman Boerhaave [18] (1668–1738) is sometimes referred to as a "father of physiology" due to his exemplary teaching in Leiden and textbook *Institutiones medicae* (1708).

It was between 1650 and 1800 that the science of modern dentistry developed. It is said that the 17th century French physician Pierre Fauchard [19] (1678–1761) started dentistry science as we know it today, and he has been named the "Father of Modern Dentistry".

Pierre Vernier [20] (1580–1637) was inventor and eponym of the vernier scale used in measuring devices. Evangelista Torricelli [21] (1607–1647) was best known for his invention of the barometer. Although Franciscus Vieta [22] (1540–1603) gave the first notation of modern algebra, John Napier [23] (1550–1617) invented logarithms, and Edmund Gunter [24] (1581–1626) created the logarithmic scales (lines, or rules) upon which slide rules are based. It was William Oughtred [25] (1575–1660) who first used two such scales sliding by one another to perform direct multiplication and division; and thus is credited as the inventor of the slide rule in 1622.

Blaise Pascal [26] (1623–1662) invented the mechanical calculator in 1642. The introduction of his Pascaline [27] in 1645 launched the development of mechanical calculators first in Europe and then all over the world. He also made important contributions to the study of fluid and clarified the concepts of pressure and vacuum by generalizing the work of Evangelista Torricelli. He wrote a significant treatise on the subject of projective geometry at the age of sixteen, and later corresponded with Pierre de Fermat [28] (1601–1665) on probability theory, strongly influencing the development of modern economics and social science.

Gottfried Leibniz [29] (1646–1716), building on Pascal's work, became one of the most prolific inventors in the field of mechanical calculators; he was the first to describe a pinwheel calculator in 1685 and invented the Leibniz wheel, used in the arithmometer, the first mass-produced mechanical calculator. He also refined the binary number system, foundation of virtually all modern computer architectures.

John Hadley [30] (1682–1744) was mathematician inventor of the octant, the precursor to the sextant. Hadley also developed ways to make precision aspheric and parabolic objective mirrors for reflecting telescopes, building the first parabolic Newtonian telescope and a Gregorian telescope with

Unit Six   Impact of Science in Europe

accurately shaped mirrors.

Denis Papin [31] (1647–1712) was best known for his pioneering invention of the steam digester, the forerunner of the steam engine. Abraham Darby I [32] (1678–1717) was the first, and most famous, of three generations with that name in an Abraham Darby family that played an important role in the Industrial Revolution. He developed a method of producing high-grade iron in a blast furnace fueled by coke rather than charcoal. This was a major step forward in the production of iron as a raw material for the Industrial Revolution. Thomas Newcomen [33] (1664–1729) perfected a practical steam engine for pumping water, the Newcomen steam engine. Consequently, he can be regarded as a forefather of the Industrial Revolution.

In 1672, Otto von Guericke [34] (1602–1686), was the first human on record to knowingly generate electricity using a machine, and in 1729, Stephen Gray [35] (1666–1736) demonstrated that electricity could be "transmitted" through metal filaments. The first electrical storage device was invented in 1745, the so-called "Leyden jar" [36], and in 1749, Benjamin Franklin [37] (1706–1790) demonstrated that lightning was electricity. In 1698 Thomas Savery [38] (c.1650–1715) patented an early steam engine.

German scientist Georg Agricola [39] (1494–1555), known as "the father of mineralogy", published his great work *De re metallica* [40]. Robert Boyle [41] (1627–1691) was credited with the discovery of Boyle's Law [42]. He is also credited for his landmark publication *The Sceptical Chymist*, where he attempts to develop an atomic theory of matter. The person celebrated as the "father of modern chemistry" is Antoine Lavoisier [43] (1743–1794) who developed his law of Conservation of mass in 1789, also called *Lavoisier's Law*. Antoine Lavoisier proved that burning was caused by oxidation, that is, the mixing of a substance with oxygen. He also proved that diamonds were made of carbon and argued that all living processes were at their heart chemical reactions. In 1766, Henry Cavendish [44] (1731–1810) discovered hydrogen. In 1774, Joseph Priestley [45] (1733–1804) discovered oxygen.

German physician Leonhart Fuchs [46] (1501–1566) was one of the three founding fathers of botany, along with Otto Brunfels [47] (1489–1534) and Hieronymus Bock [48] (1498–1554) (also called Hieronymus Tragus). Valerius Cordus [49] (1515–1554) authored one of the greatest pharmacopoeias and one of the most celebrated herbals in history, *Dispensatorium* (1546).

In his *Systema Naturae* [50], published in 1767, Carl von Linné [51] (1707–1778) catalogued all the living creatures into a single system that defined their morphological relations to one another: the Linnaean classification system. He is often called the "Father of Taxonomy". Georges Buffon [52] (1707–1788), was perhaps the most important of Charles Darwin's [53] predecessors. From 1744 to 1788, he wrote his monumental *Histoire naturelle, générale et particulière*, which included everything known about the natural world up until that date.

Along with the inventor and microscopist Robert Hooke [54] (1635–1703), Sir Christopher Wren [55] (1632–1723) and Sir Isaac Newton (1642–1727), English scientist and astronomer Edmond Halley [56] (1656–1742) was trying to develop a mechanical explanation for planetary motion. Halley's star catalogue of 1678 was the first to contain telescopically determined locations of southern stars.

Many historians of science have seen other ancient and medieval antecedents of these ideas. It is widely accepted that Copernicus's *De revolutionibus* followed the outline and method set by Ptolemy in his *Almagest* and employed geometrical constructions that had been developed previously by the Maragheh school in his heliocentric model, and that Galileo's mathematical treatment of acceleration and his concept of impetus rejected earlier medieval analyses of motion, rejecting by name: Averroes, Avempace, Jean Buridan, and John Philoponus.

The standard theory of the history of the scientific revolution claims the 17th century was a period of revolutionary scientific changes. It is claimed that not only were there revolutionary theoretical and experimental developments, but that even more importantly, the way in which scientists worked was radically changed. An alternative anti-revolutionist view is that science as exemplified by Newton's *Principia* was anti-mechanist and highly Aristotelian, being specifically directed at the refutation of anti-Aristotelian Cartesian mechanism, and not more empirical than it already was at the beginning of the century or earlier in the works of scientists such as Benedetti, Galileo Galilei, or Johannes Kepler.

New Words and Expressions

deprecation 反对，对……表示不赞成
lend itself to 有助于，适宜于
ethereal 以太的；空气一样的
impart 给予
venous 静脉的
artery 动脉
vein 静脉
impel 推动，推进
pedigree 起源，由来；谱系
ancestry [总称]祖先；世系，血统
scholium 附注；例证
axiomatic 公理的；自明的
memo 备忘录
interminable 无休止的，没完没了的
locomotion 运动；移动；旅行
geocentric 以地球为中心的
anatomical 解剖学上的；解剖的
anatomist 解剖学者

physiology 生理学
eponym 人名名称
logarithm 对数
pinwheel [机] 针轮，销轮；快速旋转的东西
arithmometer 四则计算机（一种初期的计算机）
octant 八分仪
sextant 六分仪
aspheric 非球面的
parabolic 抛物线的
filament 细丝，细线
oxidation 氧化作用
hydrogen 氢
pharmacopoeia 药典；配药书
morphological 形态（学）的
monumental 不朽的；纪念的
microscopist 显微镜学家
refutation 驳斥，反驳
Cartesian 笛卡尔的

Notes

1. Aristotelian theory 亚里士多德理论是古希腊哲学家亚里士多德（公元前384—前322）开创的有关物理学本质的理论。

2. Earth 特指亚里士多德理论中的一种纯元素"土"，并不是指由多种元素组成的地球。句中 Water、Air、Fire 分别指"水""气""火"，都是亚里士多德理论中的纯元素。

3. impetus theory 冲力说是6世纪时学者约翰·斐劳波诺斯（John Philoponus）提出的。他否认天体由神灵推动的自然观。他认为上帝创世之初就赋予天体一种"冲力"。这是一种不随时间流逝的动力，这种动力可以维持物体永远运动下去。因此，运动的物体一般并不需要经常有个推动者和它接触。

4. Galen（130—200）盖伦，希腊解剖学家、内科医生和作家，其著作对中世纪的医学有决定性影响。

5. William Harvey（1578—1657）威廉·哈维，英国17世纪著名的生理学家和医生。早年致力于古典医学著作的研究，著有《心血运动论》，被称为全部生理学史上最重要的著作。哈维因为心血系统的出色研究（以及他的动物生殖的研究），使得他成为与哥白尼、伽利略、牛顿等人齐名的科学革命巨匠。他的《心血运动论》一书也像《天体运行论》《关于托勒密和哥白尼两大体系的对话》《自然哲学之数学原理》等著作一样，成为科学革命时期以及整个科学史上极为重要的文献。

6. Nicolaus Copernicus（1473—1543）尼古拉·哥白尼，出生于波兰。40岁时，哥白尼提出了日心说，并经过长年的观察和计算完成了伟大著作《天体运行论》。

7. Johannes Kepler（1571—1630）约翰尼斯·开普勒，德国天文学家。他提出行星运动三大定律，终结了传统的周转圆理论，开创了天文的新纪元。

8. Isaac Newton（1642—1727）艾萨克·牛顿，英国物理学家、数学家、天文学家、自然哲学家和炼金术士。他在1687年发表的专著《自然哲学的数学原理》里，对万有引力和三大运动定律进行了描述。这些描述奠定了此后三个世纪里物理世界的科学观点，并成为现代工程学的基础。他通过论证开普勒行星运动定律与他的引力理论间的一致性，展示了地面物体与天体的运动都遵循着相同的自然定律；为太阳中心说提供了强有力的理论支持，并推动了科学革命。

9. Galileo Galilei（1564—1642）伽利略，意大利物理学家、天文学家和哲学家，近代实验科学的先驱。

10. three laws of motion 牛顿运动定律是英国物理泰斗牛顿所提出的三条运动定律的总称，描述物体与力之间的关系，被誉为是经典力学的基础。

11. Christiaan Huygens（1629—1695）克里斯蒂安·惠更斯，荷兰物理学家、天文学家、数学家、钟表学家和土卫六的发现者。他还发现了猎户座大星云和土星光环。

12. Geocentric model 地心说（或称天动说）是古人认为地球是宇宙的中心、其他的星球都环绕地球运行的一种学说。古希腊的托勒密（Claudius Ptolemy）将地心说的模型发展完善，且为了解释某些行星的逆行现象（即在某些时候，从地球上看那些星体的运动轨迹，有时这些星体会往反方向行走），因此他提出了本轮的理论，即这些星体除了绕地轨道外，还会沿着一些小轨道运转。后来，天主教教会接纳它为世界观的"正统理论"。托勒密的理论能初步解释从地球上所看到的现象。但是在文艺复兴时代，随着科学技术的进步，一些支持日心说的证据逐渐出现，而且有些证据无法以地心说解释，地心说逐渐占了下风。在现代世界，支持地心说的人已经寥寥无几。

13. *De revolutionibus orbium coelestium*《天体运行论》是波兰天文学家哥白尼的天文学说著作。

14. Andreas Vesalius（1514—1564）安德烈·维萨里，著名的医生和解剖学家，近代人体解剖学的创始人。他建立的解剖学为血液循环的发现开辟了道路。

15. *De humani corporis fabrica*《人体的构造》是维萨里的主要贡献，该书总结了当时解剖学的成就。

16. Ambroise Paré（约1510—1590）昂布鲁瓦兹·帕雷，法国国王的外科医生，使用葡萄酒药膏来治疗病人的伤口。这点在人们揭示了葡萄酒含有抗生素后就更能被大众所理解。

17. Matteo Realdo Colombo（约1516—1559）科伦布，意大利解剖学

教授、外科医生。

18. Herman Boerhaave（1668—1738）赫尔曼·布尔哈夫，荷兰知名的植物学家、人文主义者和医生。他被视为临床教学以及现代医院的奠基人。他的主要成就是指明了症状与病变的关系。

19. Pierre Fauchard（1678—1761）皮埃尔，法国医生，曾出版过有关牙医外科的专著，并完善了牙科临床工作，因而被称为"现代牙科之父"。

20. Pierre Vernier（1580—1637）维尼尔·皮尔，法国数学家，1631年发明了游标卡尺。

21. Evangelista Torricelli（1607—1647）托里拆利，意大利物理学家、数学家。1641年写了第一篇论文《论自由坠落物体的运动》，发展了伽利略关于运动的想法。经推荐做了伽利略的助手，伽利略去世后接替伽利略成为宫廷数学家。托里拆利在流体力学、静力学等方面做出了贡献。

22. Franciscus Vieta（1540—1603）弗朗索瓦·韦达，16世纪法国最有影响的数学家之一。最早系统地引入代数符号，推进了方程论的发展，被称为"代数学之父"。著有《应用于三角形的数学定律》《分析方法入门》等书。韦达最早明确给出有关圆周率的无穷运算式，而且创造了一套十进分数表示法，促进了记数法的改革。之后，韦达用代数方法解决几何问题的思想由笛卡尔继承，发展成为解析几何。他的研究工作为近代数学的发展奠定了基础。

23. John Napier（1550—1617）纳皮尔，早年从事神学工作，但他对数学也有着浓厚的兴趣。他以欧几里得的方式证明了罗马教皇是反基督者以及世界的末日就在1786年。他自认为《圣约翰启示录中的一个平凡发现》一书是他最重要的贡献。

24. Edmund Gunter（1581—1626）埃德蒙·甘特，英国数学家，发明了计算、航海和测量工具，提出了"余弦"和"余切"的概念。

25. William Oughtred（1575—1660）威廉·奥特雷德，英国数学家，1622年发明了计算尺。

26. Blaise Pascal（1623—1662）布莱士·帕斯卡，法国数学家、物理学家、思想家。

27. Pascaline 加法器是法国数学家帕斯卡（Blaise Pascal）1642年发明

的第一台真正的机械计算器。全名为滚轮式加法器，其外观上有6个轮子，分别代表着个、十、百、千、万、十万等。只需要顺时针拨动轮子，就可以进行加法，而逆时针则进行减法。原理和手表很像，可以说是计算机的开山鼻祖。

28. Pierre de Fermat（1601—1665）费马，法国著名数学家，被誉为"业余数学家之王"。

29. Gottfried Leibniz（1646—1716）戈特弗里德·威廉·莱布尼茨，德国哲学家和数学家。在数学史和哲学的历史中占有重要位置。

30. John Hadley（1682—1744）约翰·哈德利，英国数学家和发明家、天才机械师。他改进并制作了用于天文学方面的、有足够精度和放大率的第一批反射望远镜。1730年，他发明了双反射八分仪，用于测量太阳或一颗恒星在地平线上的经纬度。

31. Denis Papin（1647—1712）帕潘，法国著名物理学家。他致力于蒸汽泵的实验设计，发明了带有活塞的蒸汽泵。1680年发明了安全阀。这样，第一台可以把热能转变为机械能的实验型蒸汽泵就在英国试验成功了。

32. Abraham Darby I（1678—1717）亚伯拉罕·达比，开发出以焦炭而不是木炭为燃料生产生铁的方法，在工业革命中发挥了重要的作用。

33. Thomas Newcomen（1664—1729）汤玛斯·纽科门，英国工程师和发明家。曾发明过纽科门蒸汽引擎，后来被运用在矿区与油田，节省了大量的人力。

34. Otto von Guericke（1602—1686）奥托·冯·格里克，德国物理学家、政治家。

35. Stephen Gray（1666—1736）斯蒂芬·格雷，英国发明家，因发现电传导现象而被称为"电之父"。

36. Leyden jar 莱顿瓶是荷兰电学家马森布罗克（Peter Van Musschenbroek，1692—1761）发明的用来储存电流的装置，因他在荷兰莱顿大学任教，这种蓄电器后来便称为莱顿瓶。

37. Benjamin Franklin（1706—1790）本杰明·富兰克林，美国著名政治家、科学家、出版商、印刷商、记者、作家、慈善家；更是杰出的外交家及发明家。他曾经进行多项关于电的实验，并且发明了避雷针。他还发明了双焦点眼镜、蛙鞋等。

38. Thomas Savery（c.1650—1715）托马斯·萨弗里，制造了第一台蒸汽发动机。

39. Georg Agricola（1494—1555）格奥尔格·阿格里柯拉，德国自然科学家和现代矿物学的奠基人。

40. *De re metallica*《论矿冶》是德国矿冶学家阿格里柯拉（Georgius Agricola）于1550年撰写的巨著，论述了当时欧洲矿藏的开采和冶炼的技术。

41. Robert Boyle（1627—1691）罗伯特·波义耳，爱尔兰自然哲学家，在化学和物理学研究上都有杰出贡献。虽然他的化学研究仍然带有炼金术色彩，但他的《怀疑派的化学家》一书仍然被视作化学史上的里程碑。

42. Boyle's Law 波义耳定律，即在定量定温下，理想气体的体积与气体的压力成反比。

43. Antoine Lavoisier（1743—1794）安东尼·拉瓦锡。18世纪下半期，拉瓦锡发展出一套崭新的理论，因而改变了化学这门科学。他的贡献为他赢得了近代化学奠基者的封号。

44. Henry Cavendish（1731—1810）亨利·卡文迪许，英国化学家、物理学家。

45. Joseph Priestley（1733—1804）约瑟夫·普里斯特利，发现氧气的英国化学家。

46. Leonhart Fuchs（1501—1566）莱昂哈德·福克斯，德国医生。他的著作（*De historia stirpium*）是那个时代最优秀的植物学著作。他与奥托（Otto Brunfels）和海欧纳莫斯·博克（Hieronymus Bock）被称为"德国的植物学之父"。

47. Otto Brunfels（1489—1534）奥托，德国最早的植物学家。

48. Hieronymus Bock（1498—1554）海欧纳莫斯·博克，德国植物学家。他的著作（*Neu Kreunterbuck*）非常细致地描述了植物，是分类系统的开端。

49. Valerius Cordus（1515—1554）考尔都斯，德国植物学家。他的著作

Unit Six　Impact of Science in Europe

《植物历史》(*Historia plantarum*)描述了446种被子植物，而且其植物形态描述都是从活的植物直接观察得到的。

50. *Systema Naturae*《自然系统》是卡尔·冯·林奈的主要著作之一。

51. Carl von Linné [Carolus Linnaeus]（1707—1778）卡尔·冯·林奈，瑞典动植物分类学家。1735年发表了最重要的著作《自然系统》，1737年出版《植物属志》，1753年出版《植物种志》，建立了动植物命名的双名法，对动植物分类研究的进展有很大的影响。

52. Georges Buffon（1707—1788）布丰伯爵，法国博物学家、数学家、生物学家和启蒙时代著名作家。布丰的思想影响了之后两代的博物学家，包括达尔文和拉马克。布丰在博物学上的作品包括《论自然史的研究方法》《地球论》《动物史》和《自然通史》。布丰的作品对现代生态学的影响深远。

53. Charles Darwin（1809—1882）查尔斯·罗伯特·达尔文，英国生物学家和博物学家。达尔文早期因地质学研究而著名，而后又提出科学证据，证明所有生物物种是由少数共同祖先，经过长时间的自然选择过程后演化而成。到20世纪30年代，达尔文的理论成为对演化机制的主要诠释，并成为现代演化思想的基础，在科学上可对生物多样性进行一致且合理的解释，是现今生物学的基石。

54. Robert Hooke（1635—1703）罗伯特·胡克，英国博物学家和发明家。在物理学研究方面，他提出了描述材料弹性的基本定律——胡克定律，且提出了万有引力的平方反比关系。在机械制造方面，他设计制造了真空泵、显微镜和望远镜等。

55. Sir Christopher Wren（1632—1723）克里斯多佛·雷恩爵士，英国天文学家和建筑师。

56. Edmond Halley（1656—1742）埃德蒙多·哈雷，英国天文学家、地理学家、数学家、气象学家和物理学家。最著名的成就是计算出哈雷彗星的公转轨道，并正确预言了哈雷彗星作回归运动的事实。

Exercises

I. Find out the English equivalents to the following Chinese terms from the passage.

1. 日心说
2. 冲力说
3. 直线运动
4. 静脉和动脉系统
5. 三大运动定律
6. 万有引力定律
7. 地心说
8. 人体解剖学
9. 循环系统
10. 游标卡尺
11. 气压计
12. 计算尺
13. 加法器
14. 射影几何学
15. 概率论
16. 四则运算机
17. 二进制数
18. 格雷果里反射式望远镜
19. 莱顿瓶（蓄电器）
20. 质量守恒定律

II. Choose the best answer according to the information in the passage.

1. Which idea doesn't belong to the new ideas contributed to what is called the scientific revolution?

    A) The Sun is the center of the universe.

    B) Matter is continuous and made up of the elements Earth, Water, Air, and Fire.

    C) All bodies are heavy and move according to the same physical laws.

    D) Blood circulates from the arteries to the veins impelled in a circle, and is in a state of ceaseless motion.

2. According to Newton himself, his first law of motion was similar to _____ .

    A) Aristotle's counterfactual principle of interminable locomotion in a void

    B) heliocentric system

    C) the Axioms Scholium of his Principia

Unit Six   Impact of Science in Europe

   D) the impetus theory that unnatural motion is caused by continuous action of the original force
3. Who is regarded as one of the "Fathers of Surgery"?
   A) Andreas Vesalius   B) William Harvey
   C) Ambroise Pare   D) Herman Boerhaave
4. Which statement is true according to the passage?
   A) Pierre Vernier was inventor of the slide rule used in measuring devices.
   B) John Napier invented the logarithmic scales.
   C) Blaise Pascal invented Pascaline in 1645.
   D) Gottfried Leibniz was the first to describe the binary number system.
5. When was the first electrical storage device invented?
   A) 1729   B) 1745   C) 1749   D) 1789

Part C Extended Reading

Science in the Age of Enlightenment

The Age of Enlightenment was a European affair. The 17th century "Age of Reason" opened the avenues to the decisive steps towards modern science, which took place during the 18th century "Age of Enlightenment". Directly based on the works of Newton, Descartes [1], Pascal and Leibniz, the way was now clear to the development of modern mathematics, physics and technology by the generation of Benjamin Franklin (1706–1790), Leonhard Euler [2] (1707–1783), Mikhail Lomonosov [3] (1711–1765) and Jean le Rond d'Alembert [4] (1717–1783), epitomized in the appearance of Denis Diderot's [5] Encyclopédie between 1751 and 1772. The impact of this process was not limited to science and technology, but affected philosophy (Immanuel Kant [6], David Hume [7]), religion (notably with the appearance of positive atheism, and the increasingly significant impact of science upon religion), and society and politics in general (Adam Smith [8], Voltaire), the French Revolution [9] of 1789 setting a bloody caesura indicating the beginning of political modernity. The early modern period is seen as a flowering of the European Renaissance, in what is often known as the Scientific Revolution, viewed as a foundation of modern science.

The scientific history of the Age of Enlightenment traces developments in science and technology during the Age of Reason, when Enlightenment ideas and ideals were being disseminated across Europe and North America. Generally, the period spans from the final days of the 16th-and-17th-century Scientific Revolution until roughly the 19th century, after the French Revolution (1789) and the Napoleonic era [10] (1799–1815). The scientific revolution saw the creation of the first scientific societies, the rise

Unit Six Impact of Science in Europe

of Copernicanism, and the displacement of Aristotelian natural philosophy and Galen's ancient medical doctrine. By the 18th century, scientific authority began to displace religious authority and the disciplines of alchemy and astrology lost scientific credibility.

While the Enlightenment cannot be pigeonholed into a specific doctrine or set of dogmas, science came to play a leading role in Enlightenment discourse and thought. Many Enlightenment writers and thinkers had backgrounds in the sciences and associated scientific advancement with the overthrow of religion and traditional authority in favor of the development of free speech and thought. Broadly speaking, Enlightenment science greatly valued empiricism and rational thought, and was embedded with the Enlightenment ideal of advancement and progress. As with most Enlightenment views, the benefits of science were not seen universally; Jean-Jacques Rousseau [11] criticized the sciences for distancing man from nature and not operating to make people happier.

Science during the Enlightenment was dominated by scientific societies and academies, which had largely replaced universities as centers of scientific research and development. Societies and academies were also the backbone of the maturation of the scientific profession. Another important development was the popularization of science among an increasingly literate population. Philosophes introduced the public to many scientific theories, most notably through the Encyclopédie and the popularization of Newtonianism by Voltaire as well as by Émilie du Châtelet [12], the French translator of Newton's *Principia*. Some historians have marked the 18th century as a drab period in the history of science; however, the century saw significant advancements in the practice of medicine, mathematics, and physics; the development of biological taxonomy; a new understanding of magnetism and electricity; and the maturation of chemistry as a discipline, which established the foundations of modern chemistry.

New Words and Expressions

avenue 道路；途径，手段
epitomize 概括
atheism 无神论
caesura 中止
displacement 取代，置换
doctrine 教条；主义；学说
astrology 占星术
pigeonhole 把……搁在分类架上；把……留在记忆中；缓办
dogma 教义，教条
overthrow 推翻，打倒
backbone 脊柱；支柱，主要成分；骨干
maturation 成熟
drab 单调的，乏味的

Notes

1. Descartes（1596—1650）勒内·笛卡尔，法国著名的哲学家、数学家和物理学家。他因将几何坐标体系公式化而被认为是"解析几何之父"，对现代数学的发展做出了重要的贡献。他还是西方现代哲学思想的奠基人，是近代唯物论的开拓者，且提出了"普遍怀疑"的主张。他的哲学思想深深影响了之后的几代欧洲人，开拓了所谓"欧陆理性主义"哲学。

2. Leonhard Euler（1707—1783）莱昂哈德·欧拉，瑞士数学家和物理学家，近代数学先驱之一。欧拉在数学的多个领域，包括微积分和图论都做出过重大发现。此外，他还在力学、光学和天文学等学科有突出的贡献。

3. Mikhail Lomonosov（1711—1765）米哈伊尔·瓦西里耶维奇·罗蒙诺索夫，俄国化学家、哲学家、诗人，俄国自然科学的奠基者。

4. Jean le Rond d'Alembert（1717—1783）让·勒朗·达朗贝尔，法国物理学家、数学家和天文学家。

5. Denis Diderot（1713—1784）德尼·狄德罗，法国启蒙思想家、唯物主义哲学家、无神论者和作家，百科全书派的代表。他的最大成就是主编《百科全书，或科学、艺术和工艺详解词典》（通常称为《百科全书》）。此书概括了18世纪启蒙运动的精神。恩格斯称赞他是"为了对真理和正义的热诚而献出了整个生命"的人。他也被视为是现代百科全书的奠基人。

Unit Six　Impact of Science in Europe

6. Immanuel Kant（1724—1804）伊曼努尔·康德，德国哲学家和德国古典哲学创始人。他被认为对现代欧洲有重要影响的思想家，也是启蒙运动时期主要的哲学家。

7. David Hume（1711—1776）大卫·休谟，苏格兰哲学家、经济学家和历史学家，他被视为苏格兰启蒙运动以及西方哲学历史中最重要的人物之一。

8. Adam Smith（1723—1790）亚当·史密斯，苏格兰哲学家和经济学家。他所著的《国富论》成为第一本试图阐述欧洲产业和商业发展历史的著作。

9. French Revolution（1789—1799）法国大革命，拉开了现代社会的帷幕。

10. Napoleonic era（1799—1815）拿破仑时代。拿破仑时代始于1799年的雾月政变，终结于1815年百日王朝期间滑铁卢战役的失败或几天后的二次退位。

11. Jean-Jacques Rousseau（1712—1778）让-雅克·卢梭，瑞士裔的法国思想家、哲学家、浪漫主义作家、政治理论家和作曲家。

12. Émilie du Châtelet（1706—1749）沙特莱侯爵夫人（又译夏特莱侯爵夫人），法国数学家、物理学家和哲学家。

Exercises

I. Find out the English equivalent to the following Chinese terms from the passage.

1. 理性时代
2. 启蒙运动时代
3. 政治现代化
4. 科学革命
5. 科学社团
6. 科学信誉

II. Answer the questions based on the passage.

1. When and where did the Enlightenment happen?
2. What's the impact of Enlightenment?

3. When did scientific authority start to displace religious authority?
4. What were the characteristics of many Enlightenment writers and thinkers?
5. Were universities centers of scientific research and development during the Enlightenment?

Unit Seven
Industrial Revolution

Part A　Lecture

识别段落类型

段落是文章的重要组成部分，段落的构成是按照一定的思维模式组织安排的。不同的组织安排使段落具有不同的特点，因而形成了不同的段落类型。了解和识别段落类型有助于我们把握具体事实与细节的内在关系，加深理解段落的主要思想，提高阅读的速度和效果。本单元讲座主要分析常见英语段落的类型及其特点。

1　常见的英语段落类型

段落是由句子组成的语义单位，它既是文章的组成部分，又是自成一体、相对独立的整体。英语的段落结构往往通过语义贯穿，使段落中的其他句子与主题句直接关联；同时，通过一些逻辑衔接词显现逻辑序列和句际关系。因此，英语段落条理清晰，层次结构清楚，衔接自然流畅，句与句之间有内在的逻辑关系。如：

A natural resource like coal can be lost in at least three ways—more than most people realize. **Firstly**, we can lose a natural resource by using it up or by using it far faster than it can be replenished. Thus we lose coal by burning it. **Secondly**, we can lose a natural resource by letting it be wasted, as when we allow farmland to erode. Coal can be wasted by allowing a mine to become inoperable, or by using inefficient methods of burning it. **Thirdly**, we can so mismanage the waste products of a natural resource that they pollute or destroy other natural resources. The draining of coal waste into a freshwater river would harm wild life as well as needed supplies of pure water. We might even go a **fourth** step, to say that human

Unit Seven　Industrial Revolution

labor is also a natural resource, which can be lost by exhaustion, misuse, or no use. Thus the Pogatab Creek flood reduced this resource by putting the inhabitants out of work.

分析：本段段落层次结构非常清楚：第 1 句概括主题，后面的句子列举了造成自然资源浪费的四种原因，每一种原因都用两句话叙述，第一句概括论述，第二句是具体例证。此外，段落中使用的逻辑衔接词（firstly, secondly, thirdly, fourth 等）使句与句之间衔接成为一个有机整体。根据以上特点，我们知道本段落是使用列举法展开的。

英语中段落的展开方式很多，常见的段落类型有：时间顺序型、空间位置型、例证型、比较与对照型、因果型、定义型、列举型、分类型、次重型和混合型等。

1·1　时间顺序型

时间顺序型段落按时间顺序组织，有的由远而近，有的由近而远，但时间具有连贯性和清晰性，常用于叙述事件发生、发展与结束的过程。这种段落内句与句之间的粘合主要是依靠表示时间的词语，如时间副词、介词短语和时间状语从句等。如：

The first important discovery was made by Maxwell. He introduced the theory of electromagnetic energy. In 1864 Maxwell published a paper which suggested that light was part of the electromagnetic spectrum. In his paper he showed that when an electromagnetic force is changed, it travels through space as a wave. Maxwell, however, did not make electromagnetic waves, experiment with them, or suggest using them for communication. In 1888 Hertz proved Maxwell's theory. He transmitted the first electromagnetic wave across a room, and he received the wave signal on a simple receiver. These waves were called Hertzian waves. Hertz did not think he could use his apparatus for communication. Marconi, however, did believe that it was possible to use Hertzian waves for communication.

分析：本段依时间先后，简洁明了地介绍了从电磁能理论的创立到电磁波

（赫兹电波）及其在通信应用中三位科学家各自的贡献。

1.2 空间位置型

空间位置型段落主要包括事物内外部的状态、构成、性质、运动过程和方法等方面的描述。一般分为上、下、前、后、左、右，由表及里，由里及外，按空间顺序描述。在这种段落中，句与句的粘合手段为表示地点的词语，主要是地点状语（表示地点的副词、介词短语和从句）。如：

(1) Gulls as they soar are not always searching for food but merely having fun on the wing. (2) A gull flying along the shore is taking advantage of wind current formed when the sea air strikes the warm land and rises. (3) Gulls also love the lower, weaker air currents that form about three feet above the waves. (4) They ride them for hours, tipping from one current to the next. (5) Most people think that gulls ride behind boats for the food. (6) But garbage is only a small part of what they eat. (7) Clams and fish make up most of their diet. (8) The sea gulls follow boats for the ride—on the thermal currents the ships create at sea.

分析：这是一篇介绍海鸥习性的文章。第 1 句为主题句，其余部分通过海鸥飞翔在三个不同空间位置（along the shore, above the waves, behind boats 等），从以下三个方面阐述了主题，见表 7-1。

表 7-1　空间位置的描述

文中关键句编号	海鸥飞翔位置	海鸥的喜爱
2	along the shore	wind current
3	above the waves	lower, weaker air current
8	behind boats	thermal current

1.3 例证型

例证型段落是用实例来说明作者提出的观点，也就是我们常说的"摆事实，讲道理"。这种段落是通过一个主题句和几个说明主题句的具体实例构成。段的第一句或第二句为主题句，之后是阐释主题句的例证。常用的逻辑衔接词有 for example、for instance、take the example of、in the case of、to illustrate this、such as、that is、as follows 等。下面这段文章就采用了例证的方式。

One of the mysteries of evolution is the variety of eyes that nature has provided for her creatures.[Topic sentence] Most insects have remarkably good eyesight, provided by thousands of individual eye lenses which are joined together in compound eyes. [Transition]The fly's two compound eyes cover most of its head and glitter iridescently. Although a fly cannot see as well as man, it can peer in many directions at once and is instantly aware of any nearby movement. [First example]Bees, too, can see in many directions. They have two large compound eyes on the sides of their heads as well as three simple eyes set in a triangle on the tops of their heads.[Second example] But the number of eyes a creature has is not an absolute index to its ability to see. [Transition] A starfish has an eye on the tip of each of its five arms but nevertheless is almost totally blind. [Third example] Nor is the placement of a creature's eyes as important as one might suppose. [Transition] Although most birds have eyes at the sides of their heads and theoretically cannot see directly where they are going, some, like the kingfisher, can also focus squarely in front. [Fourth example]

分析：本段首句提出主题：即大自然赋予生物各种不同功能的眼睛，这是万物进化过程中的一个奥秘。第 2 句为转折句，表明大多数昆虫的视力都很好。然后，作者先后列举了苍蝇、蜜蜂、海星及鸟类的各种眼睛及其特殊功能。这四个例子均有力地证明了作者的论点，与主题思想保持一致，有着密切的联系。由于作者在每次举例时都采用一个转折句，所以句子衔接自然，整段文章连贯协调。

1·4 比较与对照型

为了进一步论证观点，作者往往把他所论述的内容跟读者所熟悉的内容进行比较和对照，通过这两种手段更好地阐释主题，同时使读者获得更多的信息。比较是将两种事物进行比较，以使其相似之处更加突出，更加清楚明白；

对照是将两种事物的不同之处进行对比，目的在于突出和强调它们之间的差异。比较与对照是一种常见的表现方式，只要使用得当，就能产生很强的说服力。

通过比较两个事物间的共同点或相似点的段落为比较型段落。常用的表示比较的逻辑衔接词有 as、like、the same as、similarly、compare with、in the same way、compare to、in comparison with、in common 等。而通过对照找出事物的相异点来说明问题的段落为对照型段落。常用的表示对照的逻辑衔接词有 in contrast（to/with）、by contrast、on the other hand、on the contrary、different from、opposite、even though、instead of、unlike、while、whereas、otherwise 等。很多时候比较和对照并用，很难区分。在同一段落里，有时候先比较后对照，或同时并举，构成比较和对照的混合型段落。如：

Like Lincoln, William McGuffey had been born to poor parents—a father with little ambition other than to hunt, a mother who taught her children the scriptures and dreamed of their future success. [Parentage] Like Lincoln, he was tall, raw-boned, strong, and homely. [Appearance] Like Lincoln, he was obedient, responsible, and hard-working. [Character] Like Lincoln, he thirsted for knowledge and was known to walk miles to borrow any kind of book.[Ambition] Like Lincoln, he was to influence the pattern of 19th-century American thought. [Influence] But unlike Lincoln, he was not to be remembered and loved as a person long after his death. [Reputation]

分析：本段将 19 世纪美国思想家 William McGuffey 与美国总统 Lincoln 作了比较和对照。由于 William McGuffey 鲜为人知，如果作者直接对他进行介绍，可能会使读者感到陌生。作者将他与大名鼎鼎的总统相比，就能事半功倍，使前者的形象变得更加具体突出。作者从五个方面将他俩作了比较，即他们的出身、外貌、性格、抱负，以及对 19 世纪美国所产生的影响。这些都说明了他俩有许多共同之处。为了强调这一点，作者不厌其烦地一再重复 Like Lincoln。最后，他将两人的不同之处作了对照，指出他俩在名望上有着很大的差异：前者死后被人遗忘，而后者则受人爱戴，流芳百世。作者将比较与对照的手法运用得非常成功，给人留下了深刻的印象。

Unit Seven Industrial Revolution

The camera and the eye are similar in many aspects. They both need light rays in order to function. Both have a sensitive surface on which the image is formed. In the eye the image is formed on the retina（视网膜）. In the camera the image is formed on the film. As in a camera, the image on the retina is inverted.

分析：本段采用比较和对照的方法对照相机和眼睛的相同和相异之处进行阐述。它们的相同之处是：need light to function; has a sensitive surface; the image is inverted。在描述两者的相同之处时用了 similar 和 both 这样的词。两者的相异之处是：in the eye, the image is formed on the retina；而 in the camera, the image is formed on the film。通过二者相同和不同方面的描述，把我们非常熟悉的眼睛和相对来说较陌生的照相机进行比较，使我们对照相机的特点和工作原理有了更清楚的认识。

从上例可以看出，比较对照型段落通过寻找一事物与另一事物之间的相同、相似、相异或相反之处，可以帮助读者更好地理解所涉及事物的特征。

1·5 因果型

因果型段落用于揭示事态的原因或结果，帮助读者理解为什么某一件事发生或出现。阐述时或从因到果，或从果到因，或因果连锁反应。常用的表示原因的逻辑衔接词有 because、as、since、for、due to、owing to、because of、the reason for… is…、on account of、in as much as that、result from、be result of 等，表示结果的逻辑衔接词有 so、thus、hence、therefore、consequently、then、lead to、for this reason、that is why…、so that、the result is that、as a result of… 等。

Nature has a number of different ways of burying and thus preserving the past. One way is by volcanic eruption, the most famous example is probably the great Roman town of Pompeii in southern Italy. The town was so quickly and completely covered with volcanic ashes from Mount Vesuvius one August day in 79 A.D. that the people of the town hardly knew what was happening to them. A second way may be seen in another part of Europe, where the surface of the land in the valleys of the Alps sank hundreds of years ago and was covered with water and peat. This cover preserved the wooden houses, utensils, and weapons of the prehistoric Alpine lake dwellers. In Egypt, drifting dry sands have preserved fragile records for thousands of years. In Central America, the jungle buried and preserved some of the ancient cities of prehistoric Indian peoples.

分析：本段首句是主题句：大自然有几种方式可以埋藏并保存历史。在阐述这一主题时，作者采用了因果关系法，逐一列举了大自然不同的运动方式所造成的结果，很好地阐明了主题，本段的因果关系见表7-2。

表 7-2　因果关系的描述

因	果
大自然的运动	被埋藏的地区
火山喷发	意大利南部城市庞贝
陆地变河海	阿尔卑斯山山谷
流沙	埃及地区
森林覆盖	中美洲史前印第安人的古城

1·6 定义型

定义型段落的特点是对讨论的对象的概念或意义做出科学的、合理的解释，有的解释涉及其全部特征，有的涉及部分特征。其写法多以分析、描述、比较和对比等方法对定义进行陈述和解释。定义段常用 be defined as、refer to、be referred to as、mean、be described as、be known as、be called、be thought of as 等信号词。如：

Wind is the movement of the air over the earth's surface. This movement is related to air pressure. Near the earth's surface the pressure is greater at the poles than at the equator. Away from the surface it is lower at the poles. Because of the pressure differences there is a constant flow of air. This is the main cause of wind. Winds are also deflected by the rotation of the earth. This deflection is called the Coriolis force.

分析：本段对 wind 和 Coriolis force 两个概念做了解释。什么是风？风是地球表面空气的运动，而 Coriolis force 是指由于地球的自转，风产生了转向，这种转向就称为地球偏转力或科里奥利力。这里对两个名词的解释都是采用对事物的本质内涵做简要说明的定义方式。

1·7 列举型

列举型段落可能列出某一系列的人、事情、事物或例子，也可能列出不同类别的人或事物。列举句型往往都是平行结构，常用数词（如 the first、the second 等），或副词（如 firstly、secondly、next、then、last、finally 等），或表示并列关系（如 one…、the other…、another 等）。这种罗列可将信息量较大、关系较复杂的内容安排得层次分明，条理清晰。如：

European universities and universities in the United States are different in many ways. **First**, European students enroll in fewer courses each term than United States students do. **Second**, European students seldom live at a university. Instead, they live at home and travel to classes. **Third**, most European courses are given by professors who lecture to their classes. In contrast, United States professors often ask their students questions or allow their students to form discussion groups. **Fourth**, European professors ask students to write fewer papers than United States professors do. Consequently, European students' final examinations are usually oral, whereas American students take written final examinations. **Finally**, a European university is mainly a place to study. But at most United States universities, social activities take up a large part of the students' time.

分析：本段为列举型段落。段落层次结构非常清楚，第 1 句"欧洲的大学与美国的大学有许多不同之处"是全段的主题句，后面的句子分别列举了欧洲的大学与美国的大学之间的不同之处。段落中使用逻辑衔接词（first、second、third、fourth、finally 等）使句与句之间衔接成为一个有机整体。

1.8 分类型

对人或事物根据其共性和特征进行分类的段落称为分类型段落。分类有两种类型：一种是由一般到具体，即由总类到分类；另一种是由具体到一般，即由分类到总类，说明总类的句子一般应为主题句。常见的分类型段落的信号词有 fall into (divide into)…kinds/types/sorts/classes/groups/species/ways…、include (embrace/comprise/consist of/make up of) kinds/sorts/types…、consist of、be classified as、have a few categories 等。如：

From least to most expensive, a hi-fi system consists of several basic units: a receiver (made up of a tuner, or radio section, a preamplifier, and an amplifier), a record turntable, perhaps a tape player, and a pair of speakers. Each component is constantly being refined; there is always something new in the world of hi-fi.

分析：从本段的叙述我们知道，无论是最便宜还是最昂贵的高保真系统都是由四个最基本的单位组成，即 a receiver、a record turntable、a tape player 和 a pair of speakers。而一个 receiver（接收器）又是由 tuner、preamplifier（扩音器）、amplifier（放大器）三个部分组成。分类使我们对高保真系统的构成有了清晰的认识。

1.9 次重型

有时作者想强调某一论据对于其观点的重要性。这时他就会运用次重型段落来组织安排论据。组织的方法有两种：一种是由重要、次要到不重要，一种是由次要、重要到最重要。前者叫作降级排列，后者叫作升级排列。在升级排列中，作者常用平行结构和排比句式，使意思得到逐步加强，最后达到高潮。如：

Robert Hooke's work in science was varied and important through the 1700s in England. **In the first place** he served as the head of the Royal Society, a group of the day's leading men of science; there he urged new tests and experiments to advance knowledge. He helped provide a means by which people could discuss their ideas with others who shared their concerns. Hooke was **also** a well-known architect whose advice about the design of buildings was welcome. He designed a large beautiful house in London where the British Museum now stands. Unfortunately Hooke's building burned down; the six long years of work he did on it were wiped out by a careless servant. Of all Hooke's gifts to humanity, however, **the most important** was his own research in science. He studied the movement of planets and improved tools for looking at the heavens. He invented the spiral watch spring. And, of course, it was Hooke who first described the cell as a basic part on plant tissue.

分析：霍克是促进先进的科学知识传播的学者，是广为人知的建筑家，而最重要的是在科学研究上他颇有建树。作者用 in the first place、also、the most important 等词语，层层递进地介绍了霍克多方面的才华，对段首的主题句作了阐述。

1·10 混合型

混合型段落是不同段落类型的综合应用模式。如：

(1) I wonder why American towns look so much alike that I sometimes mix them up in my memory. (2) The reference to the standard influence of mass production whose agents are the traveling salesman, the mail-order houses, the five-and-ten cent stores, the chain stores, the movies, is not sufficient. (3) If you stay two days in Bologna and in Ferrara, or in Arles and in Avignon, you will never mix them up in all your life. (4) But it may well happen that after you spend two days in St. Louis and in Kansas City the images of these towns soon merge into one. (5) I think the real reason for this is that these towns have not yet had time enough to individualize and to crystallize visible local traditions of their own. (6) Physiognomically speaking, children are much less differentiated from each other than grown people.

分析：本段为混合型段落。本段中的句 1 是主题句。句 2、句 5 是分析"美国城市看起来很相似"的原因，为因果型。句 3、句 4、句 6 为比较对照型。

科技英语阅读实用教程

2 综合练习

我们从以上对英语段落的介绍和分析中知道,每种类型的段落都有其自身的特点与作用,而且这些表现方式有时会在同一段落中交叉使用。本节我们就识别段落类型进行实践。试分析以下段落,辨别它们属于哪种段落类型。

Professional athletes in a number of fields are making more money than ever before. For example, in golf there have been three professional golfers who topped the one-million-dollar mark in lifetime earnings. In basketball, one player recently negotiated a contract providing him with an annual income of $200,000. And within the past year, several college All-Americans signed basketball contracts which made them millionaires over-night. Today the least amount that a pro-basketball player can receive annually is in excess of $13,000. Basketball's rosters contain a number of stars who receive $100,000 or more to play for a single season. Six-figure bonuses are commonplace. The same is true in football. College stars sign contracts which guarantee them financial security before they have proven themselves on the field as professionals. Well-established players receive salaries comparable to the stars in the other sports. All of the big names in the major sports profit handsomely from product endorsements, personal appearances, speaking engagements, and business opportunities made available to them. It all adds up to a highly lucrative business to those who possess the ability to participate.

分析:本段为例证型段落。段首句是主题句,概括了整段的内容。然后举了四个方面的例子,如高尔夫球、篮球和足球,对主题句加以具体说明。最后是归纳性的结束句。

The engineers who repaired the clock believe that the breakdown was the result of metal fatigue. Metals deteriorate owing to repeated stresses above a certain critical value.

Part of the clock mechanism fractured. As a result, the speed of the

Unit Seven　Industrial Revolution

gear wheels increased from about 1.5 revolutions per minute to 16,000 rpm the resulting centrifugal force threw pieces of the clock mechanism in all directions. It also damaged the mechanism which drives the clock's hands.

The metal fatigue had never been noticed, as fatigue cracks are invisible to the naked. The engineers have now fitted a device that will prevent the accidents from happening again. Any increase in speed will cause a brake to be applied to the gear wheel. The engineers are confident that Big Ben will remain accurate and reliable for another 200 years.

分析：这是一篇很典型的使用因果的手法来解释某种现象的文章。第一段说明大本钟出毛病的原因是由于金属的疲乏损坏，而金属的损坏又是由于持续超过某种极限的压力所造成的。第二段说明钟的部分机械装置破碎导致齿轮速度的增加，速度增加的结果是离心力把机械装置的碎片抛向四面八方，并且损害了驱动指针的装置。第三段讲述金属的疲乏磨损一直都未被注意到，其原因是磨损处的裂口肉眼看不见。本段另一组因果关系是速度的增加会导致齿轮刹车的被迫使用。

Seaweeds are also widely used as curative or preventative agents for medical problems, especially in many Asian maritime areas. Seaweed extracts are used as antibiotics, vermifuges or anthelmintics (i.e. to kill worms such as Ascaris), cough remedies, and in fever relief, wound healing and treatments for gout, gallstones, goiter, hypertension, and diarrhea. Specific seaweed compounds are also used in medical research and are commercially available, e.g. the amino acid kainic acid (KA) from the red seaweed Digenea simplex.

分析：本段为混合型段落，是列举型和例证型段落的组合。本段首先划定了海藻的两个应用范畴（治疗性或防治性制剂和医学研究制剂），然后一一列举了第一个范畴内的 10 种具体用途。至于第二个应用范畴，则只举了一个典型的例子加以说明。可以说，在第一个范畴内所举出的事物数量之和基本等于这个集合内所有子项之和；而在第二个范畴内所举出的事物只起一个例子的作用，并不等于所有子项之和。故前者为列举，后者为例证。本段的参考译文：作为解决医疗问题的治疗性或防治性制剂，海藻也得到了广泛的运用。这种情况在许多亚洲的海洋地区里尤为突出。海藻的提取物被用作抗生素、驱蠕虫药或驱虫剂（即用于杀死蛔虫之类的肠道寄生虫）、止咳药，并用于退烧、愈合伤口、治疗痛风、胆石症、甲状腺肿大、高血压和腹泻。特殊的海藻化合物也可用于医学研究，市场上可以买得到。例如，从红色海藻 Digenea simplex 中提取的氨基酸 kainic acid（KA）。

本单元讲座探讨了常见英语段落的展开方式。从以上的段落分析可知，每种段落类型都有其不同的特点，我们可以通过主题句与其他句子的关系以及段落里使用的逻辑衔接词来识别不同的段落类型。

The following paragraphs illustrate some of the methods of organizing paragraphs that we have discussed. In order to become more familiar with these common patterns of development, study each paragraph and figure out the type of paragraph.

1

During the first half of the nineteenth century much thought was given to building the Panama Canal. The discovery of gold in California in 1848 brought an increased demand for a transportation link across Panama. A railroad line was completed after six years of hard labor in the swamps and jungles. Over two thousand workmen died from yellow fever and malaria. In 1881 a French organization tried to build a canal across the Isthmus. For eleven years workmen struggled against heat and disease. At least 15,000 died before the French gave up their attempts to build the canal. For years the abandoned machinery lay in the jungles. At the close of the Spanish-American War the United States bought a strip of land ten miles wide across the Isthmus. Immediate attention was given to the control of diseases. In two years yellow fever was completely eliminated. Because of the work of American medical heroes, it was possible to build the splendid Panama Canal.

Type of paragraph: _____

Unit Seven   Industrial Revolution

2

An immense land mass stretches from the eastern shores of the Atlantic Ocean halfway around the world to the western shores of the Pacific Ocean. This is something called Eurasia, but most people think of this vast area as two continents. Europe is the western part. Asia occupies the larger eastern half. The Ural Mountains, the Caspian Sea, and the Black Sea serve as the dividing line. On the east, Asia is separated from North America by the narrow Bering Strait and the wide Pacific Ocean. Lying off the southeastern shores of the Asian mainland are the many islands of the Philippines and Indonesia. These islands are also a part of Asia. Like a chain, they help to join Asia with Australia. Southwest of Asia, across the Red Sea and beyond the Arabian Sea, lies Africa. Directly south of Asia is the Indian Ocean. To the north is the Arctic Ocean.

Type of paragraph: _____

3

Pollution spoils our environment in many ways, the air we breathe, for instance, is constantly polluted by smoke and by chemicals such as carbon monoxide in the exhaust fumes of cars and other kinds of motor vehicles. For wild life, however, there are even greater dangers in the pollution of water—of rivers, for example, or lakes and seas. It kills all kinds of sea animals, including birds as well as fish and other forms of marine life. Other causes of water pollution include power stations, which release warm water into rivers. This kills the fish and plants which live there. These are only a few examples; there are many more.

Type of paragraph: _____

4

All warm-blooded animals are very helpless at first. Young birds and young bats must be taught to fly. Thousands of young seals drown every year. They never learn to swim "naturally". The mother has to take them out under her flipper and show them how. Birds sing without instruction; however, they do not sing well unless they are able to hear older members of their species. Older harvest mice build better nests than beginners. Frank Buck says that the young elephant does not seem to know at first what his trunk is for. It gets in his way and seems more of a hindrance than a help until his parents show him what to do with it. Insects seem to start life equipped with all necessary reflexes, but they seem to improve their talents with practice. Young spiders, for example, begin by making quite primitive little webs. They attain perfection in their art only after much time. Older spiders, if deprived of their spinnerets (喷丝嘴), will take to hunting.

Type of paragraph: _____

5

There are many reasons why languages change, but three major causes help illustrate the concept. Initially, various languages that started from the same parent developed their own uniqueness after groups of speakers drifted away from one another to establish isolated, independent communities. Another major cause of language change is the influence of and interaction with foreign cultures, often as a result of military conquest. A continuing cause for change is rapidly expanding technology and new systems of communication that bring all cultures and languages into closer contact, with borrowing between languages, a common phenomenon in the contemporary world. All languages change as the experiences of their speakers change.

Type of paragraph: _____

Unit Seven   Industrial Revolution

6

Fascism is probably just a word for most of you. But the reality is very much present in this country. And the fact of it dominates the world today. Each year there is less and less freedom for more and more people. Put simple, fascism is the control of the state by a single man or by an oligarchy, supported by the military and the police.

Type of paragraph: _____

Part B Reading

The Steam Engine—A New Source of Power

The Industrial Revolution was a period from 1750 to 1850 where changes in agriculture, manufacturing, mining, transportation, and technology had a profound effect on the social, economic and cultural conditions of the times. It began in the United Kingdom, and then the rest of the world. The Industrial Revolution marks a major turning point in history; the steam engine was considered to be one of the most important technologies of that time. The introduction of steam power fueled primarily by coal, wider utilization of water wheels and powered machinery underpinned the dramatic increases in production capacity. The development of all-metal machine tools in the first two decades of the 19th century facilitated the manufacture of more production machines for manufacturing in other industries. The effects spread throughout Western Europe and North America during the 19th century, eventually affecting most of the world, a process that continues as industrialisation. The impact of this change on society was enormous.

The increase in the scale of production of both iron and cotton would not have been possible without the introduction of a new source of power to replace the water-wheel. This brings us to the third major development, after cotton and iron, in the early Industrial Revolution—steam power.

A steam engine of a sort had been devised by a Greek, Hero of Alexandria [1], in the 1st century A.D., just before Ptolemy. But Hero's engine couldn't do significant amounts of work and was little more than a toy. The modern steam engine grew from two roots.

The first root was the recognition of the large forces exerted by the atmosphere, as for example in von Guericke's [2] experiments in the 17th

Unit Seven   Industrial Revolution

century. The problem was how to harness these forces.

The second root was the understanding of air pressure which came from Torricelli's [3] work on early barometers. Torricelli's work on water barometers showed that the pressure of the atmosphere could balance a column of water just over 30 feet in height. This finding explained a fact which had been known since antiquity—that traditional water pumps, which in effect make use of the pressure of the atmosphere, can only raise water through a height of about 30 feet.

One way to overcome this limitation was to use pressures greater than one atmosphere, and this was done by Thomas Savery [4] in 1698 in a steam engine designed to pump water from mines. First of all it raised water through 30 feet by utilizing the pressure of the atmosphere. Then it pumped the water through a considerably greater height by means of high pressure steam generated in an iron boiler.

In principle, the higher the steam pressure, the higher the water could be pumped. In practice Savery's engine could pump water to about 150 feet, five times higher than an atmospheric pump, at a rate of about 1 horse-power. Even then the boilers often burst or came apart at the seams. The iron technology of the time wasn't able to do any better. For this reason it is unlikely that many such engines were built!

A safer and better steam engine for pumping water was developed by Newcomen [5] in 1712. Its most prominent feature is the massive beam. The left-hand end of this beam is connected to the underground water pump. The other end of the beam is driven by the steam engine on the right, which mainly consists of a single large cylinder and a steam-raising boiler at the bottom. The piston in the cylinder is connected to the beam; the piston is at the top of the cylinder.

In the first stage of a pumping cycle the cylinder is filled with steam from the boiler. Then cold water is sprayed into the cylinder causing the steam to condense and creating a partial vacuum inside the cylinder. Air pressure then forces the piston down to the bottom of the cylinder and this movement in turn brings the right-hand end of the beam down. The beam then returns to its original position under its own weight and operates the underground water pump. This engine was safer than Savery's engine because it didn't use high pressure steam—the piston is brought up by the beam and not by steam pressure. And, by making the cylinder larger, the power of the engine could be increased considerably.

Large cast-iron cylinders from the Coalbrookdale Ironworks were used in Newcomen engines all over England in the 18th century. Their power was also increased by introducing a control rod which moved up and down with the beam. Projections on the control rod operated the valves which turned the water and steam on or off as required. This early form of automatic control enabled a whole pumping cycle to be completed every five seconds, giving a typical power of about five or six horse-power. These water-pumping engines were very successful, but they were very bulky machines and this limited their application to other tasks.

The main source of inefficiency in Newcomen's engines was the repeated alternate heating and cooling of the main cylinder, and it was James Watt[6] (1736–1819) who devised an engine which avoided this wasteful procedure. Watt was originally an instrument maker who realized the inefficiency of Newcomen's engine when he was asked to repair a model of one at the University of Glasgow. Watt's solution, which he patented in 1769, was to use two cylinders, not one. One cylinder was always kept hot and the other, which he called the condenser, always cold.

This engine has many features in common with Newcomen's engine: the boiler raising steam, the main cylinder with its piston connected to the end of the massive bean, the other end of which, drives the pump. But in Watt's engine, the partial vacuum in the main cylinder is created not by cooling the cylinder but by connecting it to the cold condenser cylinder, which in turn is pumped out by an air pump. Watt's engine also used hot steam, not cold air, to force down the piston. It was about three times as efficient as Newcomen's engine.

Watt's steam engine attracted the attention of a man called Matthew

Boulton [7] (1728–1809) who manufactured metal objects such as buttons and ornaments in a small factory in Birmingham. Boulton's factory was powered by water-wheels but these stopped turning whenever the steams dried up. So Watt's new and efficient source of continuous power interested him, and he wrote to Watt to suggest that they work together.

So in 1774 Watt moved to Birmingham, and the next year he and Boulton became partners. Together they set up a factory which produced all kinds of steam engines at the rate of about 20 a year for the next 30 years or so. It was the first great engineering works in the world. In a later Boulton and Watt engine, dating from 1788, the up and down motion of the beam was converted into a rotary motion. It was developments like this that led to the widespread use of steam engines in iron works and in the early textile factories.

These engines were not really very powerful or efficient. Watt's engines were clearly much more economical than those of Savery and Newcomen but were relatively inefficient by modern standards. But, above all, what the new steam engines did offer was a continuous source of power—not subject to the changes of the weather and seasons, like windmills and water-wheels, nor to fatigue like human or animal muscle-power. As Boulton was reputed to say to visitors to his factory: "I sell here, sir, what all the world desires to have—power". And it was an increasingly versatile source of power which could be used almost anywhere.

New Words and Expressions

underpin 加强……的基础，支持，巩固
harness 利用；治理
seam 接缝
prominent 突出的，杰出的
cylinder 汽缸
piston 活塞
spray 喷，向……喷射
condense 浓缩，液化
projection 凸出（部分），凸块
valve 阀门
bulky 笨重的，体积大的

alternate 交替的，轮流的
ornament 装饰

repute [常用作被动语态] 称为，认为
versatile 多方面适用的，通用的

Notes

1. Hero of Alexandria（10—70）亚历山大·希罗，古希腊数学家，被认为是古代最伟大的实验家。希罗发明了一种叫汽转球的蒸汽机。在他的多项发明中，最著名的是风轮，这是最早利用风能的设备。一般认为他也是原子论者，他的一些思想源自克特西比乌斯（Ctesibius）的著作。

2. von Guericke（1602—1686）奥托·冯·格里克，德国物理学家。1650 年他发明了活塞式真空泵，并利用这一发明于 1657 年设计并进行了著名的马德堡半球实验，展示了大气压的大小并推翻了之前亚里士多德提出的"自然界厌恶真空"（即自然界不存在真空）的假说。

3. Torricelli（1608—1647）托里切利，意大利数学家、物理学家、气压计原理发现者。

4. Thomas Savery 托马斯·萨弗里，他利用蒸汽排水，使蒸汽通入密闭容器，然后在容器上喷冷水，使其中的蒸汽冷凝从而产生真空。他利用这种真空从矿井抽水，又利用锅炉蒸汽将容器中的水排空。他的设备被称为"矿工之友"。

5. Thomas Newcomen（1664—1729）汤玛斯·纽科门，英国工程师、发明家、纽科门蒸汽引擎的发明者。

6. James Watt（1736—1819）詹姆斯·瓦特，英国著名的发明家、工业革命时的重要人物。1776 年制造出第一台有实用价值的蒸汽机。

7. Matthew Boulton（1728—1809）马修·博尔顿，英国制造商和工程师。

Exercises

I. Find out the English equivalents to the following Chinese terms from the passage.

1. 动力源
2. 生产规模

Unit Seven　Industrial Revolution

3. 蒸汽动力　　　　　　4. 气压表

5. 马力　　　　　　　　6. 控制棒

7. 冷凝器　　　　　　　8. 纺织工厂

9. 现代标准　　　　　　10. 万用的动力来源

II. Answer the following questions by using the sentences in the text or in your own words.

1. Why do we say Hero's engine was far from perfect?
2. What is significance of Torricelli's finding on water pressure according to the text?
3. What is the main improvement on design part of Thomas Savery's facility?
4. What is the main idea of Paragraph 6?
5. Who solved the problem of inefficiency of Newcomen's engine and how did he do that?

Part C Extended Reading

Machine Tools

The Industrial Revolution could not have developed without machine tools, for they enabled manufacturing machines to be made. They have their origins in the tools developed in the 18th century by makers of clocks and watches and scientific instrument makers to enable them to batch-produce small mechanisms. The mechanical parts of early textile machines were sometimes called "clock work" because of the metal spindles and gears they incorporated. The manufacture of textile machines drew craftsmen from these trades and is the origin of the modern engineering industry.

Machines were built by various craftsmen—carpenters made wooden framings, and smiths and turners made metal parts. A good example of how machine tools changed manufacturing took place in Birmingham, England, in 1830. The invention of a new machine by Joseph Gillott, William Mitchell and James Stephen Perry allowed mass manufacture of robust, cheap steel pen nibs; the process had been laborious and expensive. Because of the difficulty of manipulating metal and the lack of machine tools, the use of metal was kept to a minimum. Wood framing had the disadvantage of changing dimensions with temperature and humidity, and the various joints tended to rack (work loose) over time. As the Industrial Revolution progressed, machines with metal frames became more common, but they required machine tools to make them economically. Before the advent of machine tools, metal was worked manually using the basic hand tools of hammers, files, scrapers, saws and chisels. Small metal parts were readily made by this means, but for large machine parts, production was very laborious and costly.

Apart from workshop lathes used by craftsmen, the first large machine tool was the cylinder boring machine used for boring the large-diameter cylinders on early steam engines. The planing machine [1], the slotting machine and the shaping machine were developed in the first decades of the 19th century. Although the milling machine was invented at this time, it was not developed as a serious workshop tool until somewhat later in the 19th century.

Military production, as well, had a hand in the development of machine tools. Henry Maudslay [2], who trained a school of machine tool makers early in the 19th century, was employed at the Royal Arsenal, Woolwich, as a young man where he would have seen the large horse-driven wooden machines for cannon boring made and worked by the Verbruggans. He later worked for Joseph Bramah on the production of metal locks, and soon after he began working on his own. He was engaged to build the machinery for making ships' pulley blocks for the Royal Navy [3] in the Portsmouth Block Mills. These were all metal and were the first machines for mass production and making components with a degree of interchangeability. The lessons Maudslay learned about the need for stability and precision he adapted to the development of machine tools, and in his workshops he trained a generation of men to build on his work, such as Richard Roberts, Joseph Clement and Joseph Whitworth.

James Fox of Derby had a healthy export trade in machine tools for the first third of the century, as did Matthew Murray of Leeds. Roberts was a maker of high-quality machine tools and a pioneer of the use of jigs and gauges for precision workshop measurement.

New Words and Expressions

batch 分批的
spindle 轴；锭子，纺锤
incorporate （使）结合
framing 框架，结构
smith 锻工，铁匠
turner 车工
nib 笔尖
interchangeability 可交替性，互换性

Notes

1. planing machine 龙门刨床主要用于刨削大型工件，也可在工作台上装夹多个零件同时加工，是工业的母机。

2. Henry Maudslay（1771—1831）亨利·莫兹利，英国机械发明家，被称为"英国机床工业之父"。

3. the Royal Navy 皇家海军是英国三军中最古老的军种。从大约 1692 年到第二次世界大战之间，皇家海军曾是世界上最大、最强的海军；并帮助英国成为 18 和 19 世纪最强盛的军事及经济强国；也是把大英帝国的影响力投射至全世界的重要工具。

Exercises

I. Find out the English equivalents to the following Chinese terms from the passage.

1. 车床　　　　　　　　2. 镗床
3. 铣床　　　　　　　　4. 龙门刨床
5. 立式刨床　　　　　　6. 牛头刨床
7. 接合点　　　　　　　8. 滑轮组
9. 夹具　　　　　　　　10. 量具

II. Choose the best answer according to the information in the passage.

1. Why did people call the parts of early machines "clock work"?

   A) Because they looked like clock.

   B) Because their working principles were very precisely.

   C) Because their work may save time.

   D) Because the metal spindles and gears were incorporated in them.

2. Before Industrial Revolution, why did people seldom use metal in machine making?

   A) Because metal was very expensive.

B) Because it was short in metal storage.

C) Because it was difficult to manipulate metal and the lack of machine tools.

D) Because it was hard to make metal bent as will.

3. Which one was not the disadvantage of wood framing?

A) It was hard to carve.

B) The dimension changed with temperature.

C) The dimension changed with humidity.

D) The joints were easy to rack.

4. When had the milling machine developed as a serious workshop tool?

A) In the 19th century.

B) In the Industrial Revolution.

C) After the Industrial Revolution.

D) In the 20th century.

5. In Maudslay's mind, what properties were important in the development of machine tools?

A) Efficiency and mass production.

B) Low-cost and automation.

C) Stability and precision.

D) Being portable and strong.

Unit Eight

Modern Science and Technology (I)

Part A　Lecture

了解衔接与连贯手段的使用

我们知道，句子是一个语法分析的理想单位。但在实际交往中，语言的基本单位是语篇，不是句子。句子和语篇的关系是"实现"关系，也就是说，语篇是依靠句子来实现的。当然，语篇不可能是句子的任意堆砌。构成语篇的句子必须在意义上和结构上相关。只有语义连贯并有所衔接的语句才能表达符合逻辑的思想，传达内涵完整的信息。而衔接和连贯就是语句成篇的重要保证。因此，了解语篇的衔接和连贯手段的使用，对我们理解语篇在句子层次和段落层次上的组织结构和连接手段，掌握语篇发展的基本规律，都有重要的意义。

1　语篇的衔接和连贯

衔接和连贯是语篇的重要特征。语篇衔接涉及语篇中的句际关系，通过词汇和语法手段使文脉相通，形成语篇的有形网络。语篇连贯是通过句际的语义关系和功能关系实现的，是语篇的无形网络。下面让我们一起了解语篇中的衔接和连贯手段。

1.1　衔接

衔接，或称词语连接，是指语篇内句子之间在语法或词汇方面有联系或两方面都有联系。通过一定的衔接手段，语篇内句子之间构成一个完整或相对完整的语义单位，体现语篇结构上的粘着性和意义上的连贯性。衔接手段多种多样，常见的有语法衔接手段和词汇衔接手段两大类。

Unit Eight Modern Science and Technology (I)

▶ *1.1.1* **语法衔接手段**

句际之间的衔接可以通过语法手段予以实现。语法手段有照应、替代、省略等。

（1）通过照应关系衔接

照应（reference）是表示语义关系的一种语法手段，是语篇中一个语言成分与另一个可以与之相互解释的成分之间的关系。照应使语篇更显其结构上的衔接和语义上的连贯。如：

Everyone was incredulous when it was reported that he had a vocation for the priesthood. Nevertheless it was true.

分析：照应在语篇中发挥着重要的作用。虽然语篇中的照应关系是通过一定的语言手段来表达的，但指代成分与所指对象之间是通过语义联系来构成照应关系的。它使发话者运用简短的指代形式来表达上下文已经或即将提到的内容（如例 1 中 it 指代的是上文中的分句 he had a vocation for the priesthood），从而使语篇在修辞上具有言简意赅的效果。照应还使语篇在结构上更加紧凑，从而使语篇成为前后衔接的整体。

就实现照应的关系而言，照应可以分为人称照应、指示照应和比较照应三种。

John has moved to a new house.

    A. He had it built last year.

    B. His wife must be delighted with it.

    C. I don't know it was his.

分析：例 2 为人称照应，是通过人称代词的各种形式予以实现的。例中 A、B、C 三句中的任何一句都可与前一句（John has moved to a new house.）衔接，这是因为它们都可以与上文的 John 构成照应关系，起到语篇衔接的作用。

英语中人称代词及相应的限定词的使用频率较高，这就要求我们对代词的指称应有一个明晰的理解。请仔细阅读下面段落，注意代词的指代关系。

Two things are outstanding in the creation of the English system of canals, and **they** characterize all the Industrial Revolution. One is that the men who made the revolution were practical men. Like Brindley, **they** often had little education, and in fact school education as it then was could only dull an inventive mind. The grammar schools legally could only teach the classical subjects for which **they** had been founded. The universities (there were only two, at Oxford and Cambridge) also took little interest in modern or scientific studies; and **they** were closed to those who did not conform to the Church of England.

分析：文中出现四个 they，它们所指代的名词各不相同，注意译文中划线的名词词组，它们在汉译中均被还原为其所指的名词。本段参考译文：在修建英国的运河网的过程中，有两点是非常突出的，而这两点也正是整个工业革命的特点。首先，发动这场革命的都是些实干家。同布林德雷一样，他们一般都没有受过什么教育。事实上，当时那种学校教育也只能扼杀人的创造性。按规定文法学校只能讲授古典学科，这些学校的办学宗旨本来就是如此。大学（当时只有两所，一所在牛津，一所在剑桥）对现代的或科学的学科也不怎么感兴趣；这两所大学还把不信奉英国国教的人关在门外。

Measurements indicate that Jupiter has temperature of 125K. **This** implies that **the planet** emits about 1.6 times the amount of energy it receives from the sun, which would, at first, suggest that we are studying a star rather than **a planet**.

分析：本例为指示照应，是用指示代词（如 this/these, that/those）、表示时间和地点的副词（如 here/there, now/then）以及不定冠词来表示的照应关系。例 4 出现了两次 planet，前者带有定冠词，指"木星"，清楚明了；后者带有不定冠词，说明的是木星所属的类别，即"行星"。

"Ah, I can see you are a bookworm like myself. Now," he added, pointing to Mahony who was regarding us with open eyes, "he is different; he goes in for games."

分析：本例为比较照应，指用形容词和副词的比较级形式或用表示异同、相似、差别、优劣等词语来实现的照应关系。例中 he is different 的具体意

Unit Eight　Modern Science and Technology (I)

义是以上文的 you are a bookworm 为参照点的，因而 different 与上文之间构成了比较照应关系，从而使上下文紧密地衔接在一起。

比较照应有的明显，有的含蓄。含蓄的往往难以处理。如：

Computer languages may **range from** detailed low level close to that immediately understood by the particular computer, **to** the sophisticated high level which can be rendered automatically acceptable to a wide range of computers.

分析：本段的参考译文为"计算机语言有低级的，也有高级的。前者比较繁琐，很接近于特定计算机能懂的语言；后者比较复杂，适用范围广，能自动为多种计算机所接受"。如果不采用分译，把 range from...to... 译成"范围从……到……"，译文则会显得累赘不堪、脉络不清。本译文打破了原文的句子结构，使用了"前者……后者……"这种结构。这样处理，逻辑合理，语言清晰而流畅。

（2）通过替代关系衔接

替代（substitution）指的是用替代形式来取代上文中的某一成分，是一种既可避免重复又能连接上下文的手段。由于替代成分和替代对象之间的关系可以使语篇上下文紧密地联系在一起，因此语篇中的替代关系同时也构成了语篇的衔接关系。如：

From time to time the doctor came to the door for a little air and rest his eyes, which were almost smoked out of his head; and whenever he **did so**, he had a word for me.

分析：本例为从句替代。句中 did so 替代上文的 came to the door for a little air and rest his eyes。读者在理解 did so 的确切含义时，必须将 did so 与它所替代的语言成分联系起来，这样就使语篇前后承接，紧密地衔接在一起。

217

（3）通过省略关系衔接

省略（ellipsis）指的是把语言结构中的某个成分省去。它是为了避免重复，使表达简练、紧凑、清晰的一种修辞方式。作为一种句法现象，省略不仅可以避免重复、突出新信息，而且也是语篇衔接的一种重要语法手段。如：

A. Would you like to hear another verse? I know twelve more.

B. Joan brought some carnations and Catherine some sweet peas.

C. How much does it cost?—Five pounds.

分析：这三句分别是名词性省略、动词性省略和从句省略。

我们在阅读科技文献时常常发现，为了避免重复，科技英语中会出现大量省略。如：

An object is said to be hot if its temperature is much higher than that of our bodies, or (an object is said to be) cold if its temperature is much lower (than that of our bodies).

分析：在用并列连词 and、or、but 或标点符号（逗号、分号）等手段把各分句并列连接起来的并列复合句中，某些相同的成分常被省略。

▶ 1.1.2 词汇衔接手段

词汇衔接是指通过词汇的选择产生的衔接关系。词汇衔接手段主要有两种，复现关系（reiteration 也称为重申）和同现关系（collocation 也称为搭配）。

（1）通过复现关系衔接

复现关系主要是通过反复使用关键词、同义词、近义词、上义词、下义词、概括词等手段体现的。如：

Unit Eight　Modern Science and Technology (I)

(1) I turned to the ascent of the peak.

(2) A. The **ascent** is perfectly easy.

　　B. The **climb** is perfectly easy.

　　C. The **task** is perfectly easy.

　　D. The **thing** is perfectly easy.

　　E. **It** is perfectly easy.

分析：例中句 (1) 和句 (2) 之间的衔接是通过词汇手段实现的。两句之间存在着复现关系。句 (2) 中 A 的手段是重复使用关键词；句 (2) 中 B 的手段是使用同义词；句 (2)C 中的手段是使用上义词；句 (2) 中 D 的手段是使用概括词；句 (2)E 中的手段是运用代词。

同一个语言成分多次出现，从内容上总是有所关联的。有时候，正是由于重复手段的使用，段落或语篇才有可能成为一个意义完整、上下连贯的整体。如：

　　Fill the **tub** to the level required, taking care that the water does not rise above the point indicated by the red line running around the inside of the **tub**. The **tub** is designed to take a family wash of up to 7 lb. Weight of dry clothes.

　　Take the clothes from the **wash tub** and place them in the spin dryer ensuring that they are distributed evenly around the **drum**. Secure the special retaining lid on top of the **drum**.

　　Switch the spinner control lever to ON. The spinner will start and suds will be returned to the **wash tub**.

分析：这是一份洗衣机说明书，文中不断重复使用 tub 和 drum。这类科技文章重复现象很多，因为这种文章需要的是科学性、准确性，而不是灵活多变的选词。

（2）通过同现关系衔接

同现关系指的是词语在语篇中同时出现的倾向性或可能性。如围绕

thirsty 一词，人们常联想到 water、drink、beer、coke、soda water、mineral water、coffee、tea 等词，也就是说，这些词可能会在语篇中同时与 thirsty 一词出现。它们属于同一个"词汇链"。

Some of the most common methods of **inputting information** are to use **magnetic tape, disks,** and **terminals.** The computer's **input device** (which might be a **keyboard**, a **tape drive** or **disk drive**, depending on the **medium** used in inputting information) reads the information into the computer. For **outputting information**, two common devices used are **a printer** which prints the new information on paper, or **a CRT display screen** which shows the results on a TV-like screen.

分析：该短文围绕 input device 这条线，将属同一个词汇链的 magnetic tape、disks、terminals、keyboard、tape drive、disk drive、medium、printer、CRT display screen 等词紧紧串在一起，因此在语篇中起到衔接上下文的作用。

反义词也常用来构成词语之间的同现关系。如：

Why does this little boy wriggle all the time? Girls don't wriggle.

分析：这两句话之所以能连在一起，是因为 boy 和 girl 成互补关系（complementarity）。由于它们的语义关系十分密切，往往同时出现，所以具有联系句子的作用。类似的词还有：stand up, sit down; like, hate; wet, dry; dollar, cent; north, south; bee, honey; king, crown; sunshine, cloud 等。这种词汇的同现关系是构成语篇的衔接手段之一。

1·2 连贯

连贯是指以信息发出者和接受者双方共同了解的情景为基础，通过推理来达到语义的连贯。语义连贯是构成话语的重要标志。语篇要做到整体结构连贯，不仅需要句子做到衔接自然，关系密切，而且还要合乎逻辑。

Unit Eight　Modern Science and Technology (I)

▶ 1.2.1 通过逻辑衔接词实现连贯

句子与句子的语义连贯可以通过表示各种关系的逻辑衔接词实现。如：

With few exceptions, **however**, this is not the case with independent voltage and current sources. **Although** an actual battery can often be thought of as an ideal voltage source, other non-ideal independent sources are approximated by a combination of circuit elements. Among these elements are the dependent sources, which are not discrete components as are many resistors and batteries **but** are in a sense part of electronic devices like transistors and operational amplifiers. **But** don't try to peel open a transistor's metal so that you can see a little diamond-shaped object. The dependent source is a theoretical element that is used to help describe or model the behavior of various electrical devices.

分析：本段落句间的语义连贯主要通过使用表示转折的逻辑衔接词（如 however、although 和 but 等）达到的，所以句子衔接自然，整段文章连贯协调。因此，理解逻辑衔接词的语义关系能帮助理解文章的发展脉络，理解文章的语篇框架，从而理解文章的主要内容。

▶ 1.2.2 通过语段内的逻辑关系连贯

段落是按一定逻辑组织构成的，根据行文结构，通过逻辑关系理解段落。

In an age of supersonic airliners it is difficult to realize that **at the beginning of the twentieth century** no one had ever flown in an aeroplane. However, people were flying in balloons and airships. The airship was based on the principle of the semi-rigid structure. **In 1900** Ferdinand von Zeppelin fitted a petrol engine to a rigid balloon. This craft was the first really successful steerable airship. **In 1919** an airship first carried passengers across the Atlantic, and **in 1929** one traveled round the world. During this time the design of airships was constantly being improved and **up to 1937** they carried thousands of passengers on regular transatlantic services for millions of miles.

分析：本段作者用时间顺序描述飞船的发展史。整个段落使用了很多时间状语，如 at the beginning of the twentieth century、in 1900、in 1919、in 1929 和 up to 1937 等，从而形成非常清楚的飞船发展的时间框架。段落中的时间状语将各句衔接起来，使段落的组织环环相扣，紧密和谐，反映了原文的脉络和意境。

The Moto 1100 is a small family car. It has a small engine which is **in the front**. The engine has **a capacity of 1100 cubic centimeters**. It is a **front wheel drive** car. The gear lever is **on the floor**. There are seats for four or five people. It has four front forward gears and a reverse. It has **a maximum speed of about 130km an hour**. One advantage of this car is that it has **a very low fuel consumption**.

分析：本段描写 1100 型小汽车的特征及空间位置。阅读理解时，从段落整体考虑，根据特征和空间位置重新组织句子，更显条理性和逻辑性。

2 综合练习

以上介绍了语篇的衔接和连贯手段，下面我们将分析衔接和连贯在科技文献中的应用。科技文献强调逻辑性和条理性，使用衔接和连贯手段，从形式上显现各种组合关系，使句际之间环环相扣，这正是逻辑性和条理性的保证。试分析下面语篇中的照应手段。

Ultraviolet rays are waves similar to light which are just beyond the violet end of the visible light spectrum. **Ultraviolet rays** are sometimes called invisible light or black light because we cannot see **them** with the human eye. **Ultraviolet rays** have wavelengths between 12 and 390 millimicrons. **The rays** with wavelengths between 300 and 390 millimicrons make people's skin tan. **They** also change a chemical in the skin into Vitamin D. **Those** between 185 and 300 millimicrons are used for killing harmful bacteria.

Unit Eight Modern Science and Technology (I)

分析：为了达到语义连贯的目的，可以混合使用语法和词汇手段中的多种关系来实现段落的衔接。本段使用了语法手段中的照应关系（如 them、the rays、they 和 those 等），也使用了词汇手段中的复现关系（如反复使用关键词 ultraviolet rays 等）。衔接手段的使用使本段文字内容紧凑，条理清楚，连贯性强。

(1) Human beings have distinguished themselves from other animals, and in **doing so** ensured their survival, by the ability to observe and understand their environment and then either to adapt to that environment or to control and adapt it to their own needs. (2) The process of careful observation, perception of a pattern in the phenomena observed, followed by exploitation of **this knowledge**, as largely inspired the area of human activity known as "science". (3) It has also provided the basis for the traditional methodology of science: objective observation and description of some phenomena, the formulation of a hypothesis or hypotheses about the events observed and possible relationships among them, the use of **these** to predict future events, the verification of the hypotheses and, on this basis, the construction of a theory of some area of natural activity. (4) While this process still underlies most scientific activity, the classic "scientific method" has been criticized from a variety of perspectives. (5) To begin with, it is apparent that the "objectivity" of science and scientists strictly characterizes only the lowest order of scientific activity—observation. (6) Even **here** it is doubtful whether anyone can be a truly impartial observer of events. (7) What some one chooses to observe and the way one observes **it** must, after all, in part be a reflection of previous experience and of ideas as to what is significant. (8) Consider, for example, the different ways in which an artist and a layman look at a painting and the different reactions they have to **the same work**.

(1) In Sentence 1, "doing so" refers to _____.

   A) observing and understanding their environment

   B) adapting to their environment

   C) distinguishing themselves from other animals

   D) controlling their environment

(2) In Sentence 2, "this knowledge" refers to _____.

    A) the knowledge of the process

    B) the knowledge of observation

    C) the knowledge of perception

    D) the knowledge of the pattern in the phenomena

(3) In Sentence 3, "these" refers to _____.

    A) some phenomena

    B) the events observed

    C) the hypotheses

    D) the possible relationships between the events

(4) In Sentence 6, "here" refers to _____.

    A) in observation

    B) in the process of criticizing

    C) in the construction of the theory

    D) in the beginning

(5) In Sentence 7, "it" refers to _____.

    A) the way of observation

    B) the thing chosen for observation

    C) the previous experience

    D) a reflection of ideas

(6) In Sentence 8, "the same work" refers to _____.

    A) the observation of events

    B) the formulation of hypotheses

    C) the painting

    D) the description of phenomena

分析：仔细阅读短文后，我们很快可以找到答案：(1)C; (2)D; (3)D; (4)A; (5)B; (6)C。

以上我们介绍了语篇的衔接和连贯手段，并分析了它们在科技文献中的应用。要真正掌握衔接和连贯手段的使用，还需要我们不断实践。

Unit Eight   Modern Science and Technology (I)

Read the following passages, and then choose the best answer to each question.

1

Beavers, North America's largest rodents, appear to lead such exemplary lives that a trapper once rather romantically observed that "beavers follow close to the line of the Ten Commandments". The Ten Commandments do not mention anything about building dams, lodges, and canals; however, and the beaver's penchant for **doing so** has got it into a lot of **hot water** lately. Fishing enthusiasts in the Midwest and New England are complaining about beaver dams that spoil streams for trout and, in the Southeast, lumber companies object whenever the animals flood out valuable stands of commercial timber. But some beaver experts champion a more charitable view. Historically, they say, this creature's impact on the environment has been tremendously significant, and its potential as a practical conservation resource is receiving more and more attention.

When it comes to modifying the landscape in a major way, the beaver ranks second only to humans among all living creatures. "Some people think of the beaver the same way they think of the gypsy moth," said one scientist. They think it just comes through and eats and destroys. What they don't understand is the fact that for creatures this animal has controlled the character of the forests and streams that **it** occupies.

(1) Which of the following is the best title for this passage?

  A) The Controversy over Beavers and the Environment

  B) New England's Beaver Population

  C) The Influence of Beavers on the Fishing Industry

  D) Beavers and the Ten Commandments

(2) In Sentence 2, the author refers to "hot water" to indicate that beavers _____.

   A) are able to cook their food   B) are in trouble

   C) have a form of plumbing   D) enjoy hot baths

(3) From the passage, which of the following can be inferred about gypsy moths?

   A) They conserve resources.   B) They build small dams.

   C) They have a bad reputation.   D) They eat fish.

(4) In the last sentence, what does the word "it" refer to?

   A) a fact   B) a century

   C) an animal   D) a character

(5) In Sentence 2, the word "doing so" refers to _____.

   A) following close to the line

   B) leading exemplary lives

   C) building dams, lodges and canals

   D) flooding out valuable stands of timber

2

Trees have a spectacular survival record. Over a period of more than 400 million years, they have evolved as the tallest, most massive and longest-lived organisms ever to inhabit the Earth. Yet trees lack a mean of defense that almost every animal has: trees cannot move away from destructive forces. Because they cannot move all types of living and nonliving enemies, fire, storms, microorganisms, insects, other animals and, later, humans have wounded them through their history. Trees have survived because their evolution has made them into highly compartmented organisms: that is, **they** wall off injured and infected wood.

Unit Eight Modern Science and Technology (I)

In that respect trees are radically different from animals. Fundamentally, animals heal: they preserve their life by making billions of repairs, installing new cells or rejuvenated cells in the positions of old **ones**. Trees cannot heal: they make no repairs. Instead, they defend themselves from the consequences of injury and infection by walling off the damage. At the same time they put new cells in new positions in effect; they grow a new tree over the old one every year. The most obvious of the process are growth rings, which are visible on the cross section of a trunk, a root, or a branch.

(1) What does the passage mainly about?

　　A) The growth rings of trees.

　　B) The survival of trees.

　　C) The living and non-living enemies of trees.

　　D) The radical difference between trees and animals.

(2) The author describes trees as all of the following EXCEPT _____.

　　A) tall　　　　B) green　　　C) massive　　D) long-lived

(3) The author implies that almost every animal is able to protect itself from destructive forces by doing which of the following?

　　A) Moving away　　　　　B) Calling for help

　　C) Climbing up a tree　　　D) Remaining with its group

(4) In the 2nd sentence of Paragraph 2, the word "ones" refers to _____.

　　A) animals　　B) trees　　　C) cells　　　D) repairs

(5) In the last sentence of Paragraph 1, the word "they" refers to _____.

　　A) enemies　　B) animals　　C) humans　　D) trees

Part B Reading

Science and Technology in the Electrical Age[1]

The capture and control of electricity by humans may be equated to human mastery of fire so many millennia ago, and the electric power industry was utterly revolutionary as well. It required a novel infrastructure of generators, transformers, and wires that came to extend across continents and under the seas and oceans. Within a few years electricity replaced gas lighting, which had itself been called into being by urbanization and industrialization. In addition to light, electricity began to provide power and heat wherever wires could reach—which meant literally around the globe. By 1900, the stage was set for electrifification by the invention of techniques for the production and distribution of electric current—the battery, the dynamo, and the development of a copper wire industry.

The 1890s and the first decades of the 20th century saw the introduction of some extremely popular new consumer items. A whole list of products changed the day-to-day life of the average person: safety razors, aspirin, thermos bottles, electric blankets, breakfast foods, cellophane, and rayon. To this array were added a constant flood of improvements to earlier inventions that also found their way into daily life, such as the camera, the typewriter, the phonograph, the refrigerator, and most of all, the automobile. The automobile, a novelty in 1895, was widespread by 1910, and it led to gas stations, traffic lights, and highway systems. The social impact was profound and dramatic. By the 1930s, the automobile had changed the commercial landscape, with motels and highway cabins, roadside diners, fruit stands, drive-in theaters, and the beginnings of coast-to-coast highway travel.

As the generation born in the 1880s and 1890s grew to adulthood,

Unit Eight   Modern Science and Technology (I)

they were stunned by the rapid changes, particularly impressed by the automobile, the radio, the airplane, and plastics. Like their grandparents, who had witnessed profound 19th-century transformations brought on by the Industrial Revolution, the first 20th-century generation frequently remarked on the positives and negatives of technical progress.

Life was more comfortable but faster. Cities became gradually cleared of the flood of horse manure, but soon the streets were choked with cars and their fumes. Automobile accidents began to claim hundreds, then thousands, of victims. Information flowed more quickly across the cable lines and later over the radio, but much of the news was itself depressing, bringing the horrors of war, political tyranny, revolution, and natural disaster to the morning breakfast table. While eating corn flakes or shredded wheat, the average citizen learned of the Japanese defeat of Russia in 1905, the progress of a naval armaments race between Britain and Germany unlike any the world had ever seen, and then the horrors of trench warfare in World War I. Even before radio was introduced, phone companies experimented with using telephone lines to carry news, speeches, and entertainment over the networks as a type of broadcast system.

The frontiers of scientific discovery moved into new areas of physics and chemistry in the 1890s and the first years of the 20th century, opening some new and startling possibilities. While Einstein's theory of relativity represented a theoretical explanation for the relationship of space and time, rather than the discovery of new laws of nature, related findings such as Planck's constant, the Heisenberg principle, and the nature of the atom and its subatomic particles suggested that some of the certainties of Newtonian physics were no longer valid.

As in the past, the pathway between invented technological tool and discovered scientific principle was a two-way street. The Crookes tube, which had tested the effect of electrical leads in a glass-enclosed vacuum in the 1880s, spun off into a number of parlor-demonstration devices. Then suddenly, in the 1890s, a burst of scientific discoveries poured out of experimentation with the Crookes tube toy-turned-tool. Within a decade, using cathode ray tube variations on Crookes's invention, scientists discovered radioactivity, radium, polonium, X-rays, and the electron. Such new discoveries in the realm of physics, including a deeper understanding of the nature of isotopes, the electron, and the nucleus of the atom, began to pave the way for the Atomic and Electronic Ages that would follow in the later years of the 20th century. And perhaps even more important for the average person, the original Crookes cathode ray tube, with some modification, provided the ideal television screen, already in experimental commercial use by the early 1930s.

The short period from the 1890s to the 1930s was one of great inventiveness, but many of the inventions that had the greatest impact on daily life were improvements over existing machines (as with electrical lighting and internal-combustion engines) or simply the ingenious creations of people with good ideas, such as the work that led to the zipper, the brassiere, and the parking meter. Such useful, everyday items could have been made decades earlier with existing tools and materials, but they had to await the matching of need, idea, and the perseverance of individual inventors.

The times had changed, and new terms to define the era seemed needed. In many ways this was an Electrical Age, with the wider spread of such electric-based inventions as the telephone, the phonograph, the electric light, the electric refrigerator, and other household appliances. As Edison had realized, a network of electrical supply systems was needed to make electric light viable, and then with the networks, vast new markets for other devices quickly followed for the office, home, and factory. Many of the new products were already available in quite workable form in 1890, but with readily available electric power, their manufacture and sale became big business, with General Electric and Westinghouse leading the way in the United States and Siemens in Germany.

Unit Eight Modern Science and Technology (I)

New weapons and ships and vehicles to carry them gave the wars of the early 20th century a different technological character than those of the 19th century, as war moved off the surface of the planet above to the air and beneath to the sea. Navies adopted the submarine, destroyer, and battleship. Radio and sonar brought early electronic devices to warfare. With older inventions, such as the machine gun and barbed wire, World War I (1914–1918) was a chaos of human slaughter. The tank, dirigible, and airplane were used in that war to devastating effect.

The new technologies of war struck deep in social impact. World War I nearly destroyed a whole generation of Europeans, leaving the survivors as the "Lost Generation", disillusioned, bitter, and alienated. Out of the alienation came a burst of creativity and modernism in art, literature, and music.

Ongoing technological changes gave the cultural products of the Lost Generation specific qualities. Employment of women in office and sales work and their enfranchisement in Britain and the United States signaled liberation, represented by changes in clothing and hairstyles and noticeably different behavior and manners. Well-brought-up young ladies were now seen smoking cigarettes in public, driving their own cars, and wearing types of clothing that would have shocked their mothers two or three decades earlier.

Technology had played a role in bringing those social changes and would continue to shape the ways in which the changes were felt. The Jazz Age denoted the rise of a new music and a new culture that went with it. In the 1920s, a young generation of people of all races began to identify with the music of the alienated or marginalized population of African Americans. The blues, deeply rooted in black culture, moved into the mainstream, not by accident, but because the music resonated with the national and international mood of youth. The phonograph record, the radio, and motion pictures all caught the changing tempo of the times and reflected it from the masses to the creative elites and back to the masses.

Music moved out of the parlor and the amateur gathering as the radio and record industries created the 20th-century phenomenon of professional popular music.

Politics and international affairs also were reshaped by technology. The cheap newspaper, set on linotype machines, with telephoto (or facsimile) pictures from all over the world, carried the news, advertising, and propaganda of the day, as did the motion picture and radio broadcasting. Mass politics still incorporated some earlier methods of arousing popular support, such as the parade, the torchlight gathering, and the public address to huge crowds, but all such events were made more dramatic with the new technologies of floodlighting, recorded music, and the amplified sound of loudspeakers. In the United States, master politicians such as Franklin Roosevelt and religious leaders such as Billy Sunday and Father Charles Coughlin became adept at using the radio to convey their messages.

Some of the popular heroes of the day were men and women who contributed to the new technologies or used them in new and striking ways. Thomas Edison, Henry Ford, and the Wright brothers became heroes of invention; Madame Curie and Albert Einstein earned fame as scientific geniuses; Charles Lindbergh and Amelia Earhart, as aviation greats. Radio, film, and recorded music created flocks of new media celebrities, sports and film idols, crooners and divas, including Jack Benny, Babe Ruth, Charlie Chaplin, Rudolph Valentino, Enrico Caruso, and Bessie Smith.

Despite such bright diversions and mass impacts, technology held a darker threat. By the 1930s, scientists could foresee that the weapons of the future, when they tapped into the energy of the atom, could be even more terrible than those of World War I. The Atomic Age was about to be born in the mid-1930s, with the discovery of the neutron and further experimentation with radiation. The science fiction and the science predictions of the 1930s became the science fact and the technological realities of the next decades.

Unit Eight　Modern Science and Technology (I)

New Words and Expressions

cellophane 玻璃纸
rayon 人造丝
manure 粪肥；肥料
choke 窒息；哽噎
shred 切碎；细条，碎片
trench 沟；战壕
parlor 客厅，起居室
polonium 钋
isotopes 同位素
ingenious 精巧的，新颖独特的
viable 可实施的，切实可行的
sonar 声纳
chaos 混乱

alienate 使疏远，离间
enfranchisement 选举权的授予
marginalize 使处于边缘，边缘化
resonate 产生共鸣；发出回响
elite 社会精英
linotype 莱纳排铸机，整行铸造排字机
celebrity 名人
idol 偶像
crooner 男歌手
diva 著名女歌唱家
tap 利用，开发，发掘（已有的资源、知识等）

Notes

1. 本文节选自 Rodney Carlisle 的 *Scientific American Inventions and Discoveries* 一书。

Exercises

I. Find out the English equivalents of the following Chinese terms from the passage.

1. 安全剃须刀
2. 保温瓶
3. 电热毯
4. 加油站
5. 商业格局
6. 海军军备竞赛
7. 普朗克常数
8. 海森堡原理
9. 克鲁克斯管
10. 玻璃封闭的真空

11. 阴极射线管
12. 电气照明
13. 内燃机
14. 停车收费表
15. 潜艇
16. 驱逐舰
17. 战舰
18. 机关枪
19. 铁丝网
20. 飞艇
21. 科幻小说
22. 科学预言

II. Translate the following sentences into Chinese.

1. By the 1930s, the automobile had changed the commercial landscape, with motels and highway cabins, roadside diners, fruit stands, drive-in theaters, and the beginnings of coast-to-coast highway travel.

2. Information flowed more quickly across the cable lines and later over the radio, but much of the news was itself depressing, bringing the horrors of war, political tyranny, revolution, and natural disaster to the morning breakfast table.

3. The frontiers of scientific discovery moved into new areas of physics and chemistry in the 1890s and the first years of the 20th century, opening some new and startling possibilities.

4. Then suddenly, in the 1890s, a burst of scientific discoveries poured out of experimentation with the Crookes tube toy-turned-tool. Within a decade, using cathode ray tube variations on Crookes's invention, scientists discovered radioactivity, radium, polonium, X-rays, and the electron. Such new discoveries in the realm of physics, including a deeper understanding of the nature of isotopes, the electron, and the nucleus of the atom, began to pave the way for the Atomic and Electronic Ages that would follow in the later years of the 20th century.

5. In many ways this was an Electrical Age, with the wider spread of such electric-based inventions as the telephone, the phonograph, the electric light, the electric refrigerator, and other household appliances.

6. Many of the new products were already available in quite workable form

in 1890, but with readily available electric power, their manufacture and sale became big business, with General Electric and Westinghouse leading the way in the United States and Siemens in Germany.

7. New weapons and ships and vehicles to carry them gave the wars of the early 20th century a different technological character than those of the 19th century, as war moved off the surface of the planet above to the air and beneath to the sea.

8. The blues, deeply rooted in black culture, moved into the mainstream, not by accident, but because the music resonated with the national and international mood of youth.

Part C Extended Reading

Development of Natural Sciences

The Scientific Revolution established science as a source for the growth of knowledge. During the 19th century, the practice of science became professionalized and institutionalized in ways that continued through the 20th century. As the role of scientific knowledge grew in society, it became incorporated with many aspects of the functioning of nation-states.

Physics

The Scientific Revolution is a convenient boundary between ancient thought and classical physics. Nicolaus Copernicus revived the heliocentric model of the solar system described by Aristarchus of Samos [1]. This was followed by the first known model of planetary motion given by Kepler in the early 17th century, which proposed that the planets follow elliptical orbits, with the Sun at one focus of the ellipse. Galileo ("Father of Modern Physics") also made use of experiments to validate physical theories, a key element of the scientific method.

In 1687, Isaac Newton published the *Principia Mathematica* [2], detailing two comprehensive and successful physical theories: Newton's laws of motion, which led to classical mechanics; and Newton's Law of Gravitation, which describes the fundamental force of gravity. The behavior of electricity and magnetism was studied by Faraday [3], Ohm [4], and others during the early 19th century. These studies led to the unification of the two phenomena into a single theory of electromagnetism, by Maxwell [5] (known as Maxwell's equations).

The beginning of the 20th century brought the start of a revolution in physics. The long-held theories of Newton were shown not to be correct

Unit Eight Modern Science and Technology (I)

in all circumstances. Beginning in 1900, Max Planck [6], Albert Einstein [7], Niels Bohr [8] and others developed quantum theories to explain various anomalous experimental results, by introducing discrete energy levels. Not only did quantum mechanics show that the laws of motion did not hold on small scales, but even more disturbingly, the theory of general relativity, proposed by Einstein in 1915, showed that the fixed background of spacetime, on which both Newtonian mechanics and special relativity depended, could not exist. In 1925, Werner Heisenberg [9] and Erwin Schrödinger [10] formulated quantum mechanics, which explained the preceding quantum theories. The observation by Edwin Hubble [11] in 1929 that the speed at which galaxies recede positively correlates with their distance, led to the understanding that the universe is expanding, and the formulation of the Big Bang [12] theory by Georges Lemaître [13].

Further developments took place during World War II, which led to the practical application of radar and the development and use of the atomic bomb. Though the process had begun with the invention of the cyclotron by Ernest O. Lawrence [14] in the 1930s, physics in the postwar period entered into a phase of what historians have called "Big Science" [15], requiring massive machines, budgets, and laboratories in order to test their theories and move into new frontiers. The primary patron of physics became state governments, who recognized that the support of "basic" research could often lead to technologies useful to both military and industrial applications. Currently, general relativity and quantum mechanics are inconsistent with each other, and efforts are underway to unify the two.

Chemistry

The history of modern chemistry can be taken to begin with the distinction of chemistry from alchemy by Robert Boyle [16] in his work *The Sceptical Chymist*, in 1661 (although the alchemical tradition continued for some time after this) and the gravimetric experimental practices of medical chemists like William Cullen [17], Joseph Black [18], Torbern Bergman [19] and

Pierre Macquer[20]. Another important step was made by Antoine Lavoisier[21] (Father of Modern Chemistry) through his recognition of oxygen and the law of conservation of mass, which refuted phlogiston theory. The theory that all matter is made of atoms, which are the smallest constituents of matter that cannot be broken down without losing the basic chemical and physical properties of that matter, was provided by John Dalton[22] in 1803, although the question took a hundred years to settle as proven. Dalton also formulated the law of mass relationships. In 1869, Dmitri Mendeleev[23] composed his periodic table of elements[24] on the basis of Dalton's discoveries.

The synthesis of urea by Friedrich Wöhler[25] opened a new research field, organic chemistry, and by the end of the 19th century, scientists were able to synthesize hundreds of organic compounds. The later part of the 19th century saw the exploitation of the Earth's petrochemicals, after the exhaustion of the oil supply from whaling. By the 20th century, systematic production of refined materials provided a ready supply of products which provided not only energy, but also synthetic materials for clothing, medicine, and everyday disposable resources. Application of the techniques of organic chemistry to living organisms resulted in physiological chemistry, the precursor to biochemistry. The 20th century also saw the integration of physics and chemistry, with chemical properties explained as the result of the electronic structure of the atom. Linus Pauling's[26] book on *The Nature of the Chemical Bond* used the principles of quantum mechanics to deduce bond angles in ever-more complicated molecules. Pauling's work culminated in the physical modeling of DNA, the secret of life (in the words of Francis Crick[27], 1953). In the same year, the Miller-Urey experiment[28] demonstrated in a simulation of primordial processes, that basic constituents of proteins, simple amino acids, could themselves be built up from simpler molecules.

Biology, medicine, and genetics

In 1847, Hungarian physician Ignác Fülöp Semmelweis[29] dramatically reduced the occurrence of puerperal fever by simply requiring physicians to wash their hands before attending to women in childbirth. This discovery predated the germ theory of disease. However, Semmelweis' findings were not appreciated by his contemporaries and came into use only with discoveries by British surgeon Joseph Lister[30], who in 1865 proved the principles of antisepsis. Lister's work was based on the important findings by French biologist Louis Pasteur[31]. Pasteur was able to link microorganisms with disease, revolutionizing medicine. He also devised one of the most important methods in preventive medicine, when in 1880 he produced a vaccine against rabies. Pasteur invented the process of pasteurization, to help prevent the spread of disease through milk and other foods.

Perhaps the most prominent, controversial and far-reaching theory in all of science has been the theory of evolution by natural selection put forward by the British naturalist Charles Darwin[32] in his book *On the Origin of Species*[33] in 1859. Darwin proposed that the features of all living things, including humans, were shaped by natural processes over long periods of time. Implications of evolution on fields outside of pure science have led to both opposition and support from different parts of society, and profoundly influenced the popular understanding of "man's place in the universe". However, Darwinian evolutionary models do not directly impact the study of genetics. In the early 20th century, the study of heredity became a major investigation after the rediscovery in 1900 of the laws of inheritance developed by the Moravian monk Gregor Mendel[34] in 1866. Mendel's laws provided the beginnings of the study of genetics, which became a major field of research for both scientific and industrial research. By 1953, James D. Watson[35], Francis Crick and Maurice Wilkins[36] clarified the basic structure of DNA, the genetic material for expressing life in all its forms. In the late 20th century, the possibilities of genetic engineering became practical for the first time, and a massive international effort began in 1990 to map out an entire human genome (the Human Genome Project[37]).

New Words and Expressions

nation-state 民族国家
elliptical 椭圆的
validate 证实
unification 统一，联合；一致
electromagnetism 电磁学
quantum 量子
anomalous 异常的，反常的
discrete 分离的，分立的；无联系的
spacetime 时空
formulate 系统地阐述成（提出）；用公式表示
preceding 先前的，在前的
recede 后退；降低，减小
cyclotron 粒子回旋加速器
frontier 尖端，新领域
patron 保护人；资助人；主顾
inconsistent (with) 不一致的，不协调的
gravimetric（测定）重量的；重量分析
phlogiston 燃素
proven 被证实的，证据确凿的
urea 尿素
exploitation 开采，发掘，利用
petrochemical 石油化学制品
disposable 可随意使用的，可（任意）处理的，易处置的
culminate 达到顶点；结束，完结
simulation 模拟，模拟试验，模拟分析
primordial 原生的，原始的，基本的
Hungarian 匈牙利的
attend to 照顾，护理
antisepsis 防腐（法），消毒（法）
microorganism 微生物
rabies 狂犬病
pasteurization 加热杀菌（法），巴斯德氏杀菌法
heredity 遗传
inheritance 遗传，继承
Moravian 摩拉维亚的
genome 染色体基因，基因组

Notes

1. Aristarchus of Samos 萨摩斯的阿利斯塔克斯，古希腊天文学家。

2. *Principia Mathematica*《自然哲学的数学原理》是英国伟大的科学家艾萨克·牛顿的代表作。成书于 1687 年。

3. Michael Faraday（1791—1867）迈克尔·法拉第，英国物理学家（当

Unit Eight   Modern Science and Technology (I)

时"物理"被称为"自然哲学"），精于化学，在电磁学及电化学领域有所贡献。

4. Ohm（1789—1854）欧姆，德国物理学家。他发现了著名的欧姆定律。电阻的国际单位制"欧姆"是以他的名字命名的。

5. Maxwell（1831—1879）麦克斯韦，英国理论物理学家和数学家、经典电动力学的创始人和统计物理学的奠基人之一。麦克斯韦被普遍认为是对20世纪最有影响力的19世纪物理学家。他对基础自然科学的贡献仅次于牛顿和爱因斯坦。

6. Max Planck（1858—1947）马克斯·普朗克，德国物理学家和量子力学的创始人，因发现能量量子而对物理学的进展做出了重要贡献。

7. Albert Einstein（1879—1955）阿尔伯特·爱因斯坦，美籍德国犹太裔、理论物理学家、相对论的创立者和现代物理学奠基人。

8. Niels Bohr（1885—1962）尼尔斯·玻尔，丹麦物理学家。他通过引入量子化条件，提出了玻尔模型来解释氢原子光谱，提出对应原理、互补原理和哥本哈根诠释来解释量子力学，对20世纪物理学的发展影响深远。

9. Werner Heisenberg（1901—1976）韦纳·海森堡，德国物理学家、量子力学的创始人之一和"哥本哈根学派"代表性人物，1932年获得了诺贝尔物理学奖。他对物理学的主要贡献是给出了量子力学的矩阵形式（矩阵力学），提出了"测不准原理"（又称"不确定性原理"）和S矩阵理论等。他的《量子论的物理学基础》是量子力学领域的一部经典著作。

10. Erwin Schrödinger（1887—1961）埃尔温·薛定谔，奥地利理论物理学家和量子力学的奠基人之一。1933年和英国物理学家狄拉克共同获得了诺贝尔物理学奖，被称为"量子物理学之父"。

11. Edwin Hubble（1889—1953）爱德温·鲍威尔·哈勃，美国著名的天文学家。他证实了银河系外其他星系的存在，并发现了大多数星系都存在红移的现象建立了哈勃定律（这是宇宙膨胀的有力证据）。哈勃是公认的星系天文学创始人和观测宇宙学的开拓者，并被天文学界尊称为"星系天文学之父"。为纪念哈勃的贡献，小行星2069、月球上的哈勃环形山以及哈勃太空望远镜均以他的名字来命名。

12. Big Bang 宇宙大爆炸是根据天文观测研究后得到的一种设想。大约在 150 亿年前，宇宙所有的物质都高度密集在一点，有着极高的温度，因而发生了巨大的爆炸。大爆炸以后，物质开始向外大膨胀，就形成了今天我们看到的宇宙。

13. Georges Lemaître（1894—1966）乔治·爱德华·勒梅特，比利时牧师、宇宙学家。他的主要成就是提出有关宇宙起源的大爆炸理论，代表作品有《论宇宙演化》和《原始原子假说》。

14. Ernest O. Lawrence（1901—1958）欧内斯特·奥兰多·劳伦斯，美国物理学家，1939 年诺贝尔物理学奖获得者。

15. Big Science（或 Mega-science，Large Science）大科学一般是指投资大、多学科交叉的大型基础科学研究项目。大科学对促进各国的基础科学研究具有很大的意义。

16. Robert Boyle（1627—1691）罗伯特·波义耳，英国化学家。

17. William Cullen（1710—1790）库伦，苏格兰物理学家、化学家和农业学家。

18. Joseph Black（1728—1799）约瑟夫·布莱克，英国化学家和物理学家。他是蒸汽机发明家瓦特的好友。他参与创立了詹姆斯·哈顿的地质学学说。

19. Torbern Bergman（1735—1784）柏格曼，瑞典化学家和矿物学家。

20. Pierre Macquer（1718—1784）马科，法国化学家。

21. Antoine Lavoisier（1743—1794）安东尼·拉瓦锡，近代化学奠基者。

22. John Dalton（1766—1844）约翰·道尔顿，英国科学家。在 19 世纪初把原子假说引入了科学主流。

23. Dmitri Mendeleev（1834—1907）德米特里·门捷列夫，俄国化学家。他发现了元素周期律，并就此发表了世界上第一份元素周期表。

24. periodic table of elements 元素周期表。化学元素周期表是根据原子序从小至大排序的化学元素列表。由于周期表能够准确地预测各种元素的特性及其之间的关系，因此在化学及其他科学范畴中被广泛使用。

25. Friedrich Wöhler（1800—1882）弗里德里希·维勒，德国化学家。他因人工合成尿素、打破有机化合物的"生命力"学说而闻名。

26. Linus Pauling（1901—1994）莱纳斯·鲍林，美国著名化学家和量

Unit Eight　Modern Science and Technology (I)

子化学和结构生物学的先驱之一。1954 年因在化学键方面的工作获得诺贝尔化学奖。鲍林被认为是 20 世纪对化学科学影响最大的科学家之一，他所撰写的《化学键的本质》被认为是化学史上最重要的著作之一。

27. Francis Crick（1916—2004）弗朗西斯·克里克，著名英国生物学家、物理学家及神经科学家。他最重要的成就是 1953 年在剑桥大学卡文迪许实验室与詹姆斯·沃森共同发现了脱氧核糖核酸（DNA）的双螺旋结构，二人也因此与莫里斯·威尔金斯共同获得了 1962 年诺贝尔生理及医学奖。

28. Miller-Urey experiment 米勒–尤列实验是一项模拟假设性早期地球环境的实验，研究目的是测试化学演化的发生情况。该学说认为早期地球环境使无机物合成有机化合物的反应较易发生。该实验是关于生命起源的经典实验之一，由芝加哥大学的史坦利·米勒与哈罗德·尤列于 1953 年主导完成。

29. Ignác Fülöp Semmelweis（1818—1865）塞麦尔维斯，匈牙利产科医师、现代产科消毒法倡导者之一。他在维也纳（1847—1849）和布达佩斯（1850）医院产科工作时，经过细致观察，证实了产褥热是由于接生人员的手或器械受到污染传染产妇引起的败血症；于是他提倡使用漂白粉溶液消毒接生人员的手和器械；后来采用这种方法的医院产褥热死亡率显著减少。

30. Joseph Lister（1827—1912）约瑟夫·李斯特，英国外科医生，外科手术消毒技术的发明者和推广者。他因对红血球的研究而当选为英国皇家学会院士。

31. Louis Pasteur（1822—1895）路易斯·巴斯德，法国微生物学家、化学家和微生物学的奠基人之一。他以否定自然发生说（自生说）并倡导疾病细菌学说（胚种学说）和发明预防接种方法而闻名，他也是第一个创造狂犬病和炭疽疫苗的科学家。他被世人称颂为"进入科学王国的最完美无缺的人"，也被视为"细菌学之祖"。

32. Charles Darwin（1809—1882）查尔斯·罗伯特·达尔文，英国生物学家和博物学家。达尔文早期因地质学研究而著名，而后又提出科学证据，证明所有生物物种是由少数共同祖先经过长时间的自然选择过程后演化而成。到了 20 世纪 30 年代，达尔文的理论成为对演化机制的主要诠释，并成为现代演化思想的基础，在科学上可对生物多样性进行一致且合理的解释，是现今生物学的基石。

33. On the Origin of Species《物种起源》是达尔文论述生物演化的重要著作，出版于 1859 年。该书是 19 世纪最具争议的著作之一，其中的大多数观点为当今的科学界普遍接受。在该书中，达尔文首次提出了演化论的观点。他使用自己在 19 世纪 30 年代环球科学考察中积累的资料，试图证明物种的演化是通过自然选择（天择）和人择的方式实现的。

34. Gregor Mendel（1822—1884）格雷戈尔·孟德尔，奥地利遗传学家、天主教圣职人员和遗传学的奠基人。

35. James D. Watson（1928— ）詹姆斯·杜威·沃森，美国分子生物学家和 20 世纪分子生物学的牵头人之一。

36. Maurice Wilkins（1916—2004）莫里斯·威尔金斯，英国分子生物学家，专注于磷光、雷达、同位素分离与 X 光衍射等领域。他在伦敦国王学院期间解开了 DNA 分子结构以及一些相关研究，因此与克里克、沃森共同获得了 1962 年的诺贝尔生理学或医学奖。

37. Human Genome Project（HGP）人类基因组计划由美国科学家于 1985 年率先提出，于 1990 年正式启动。美国、英国、法国、德国、日本和中国科学家共同参与了这一预算高达 30 亿美元的人类基因组计划。按照这个计划的设想，2005 年要把人体内约 10 万个基因的密码全部解开，同时绘制出人类基因的谱图。换句话说，就是要揭开组成人体 4 万个基因的 30 亿个碱基对的秘密。人类基因组计划、曼哈顿原子弹计划和阿波罗计划并称为"三大科学计划"。

Exercises

I. Find out the English equivalents to the following Chinese terms from the passage.

1. 行星的运行　　　　　　　　2. 万有引力定律

3. 量子力学　　　　　　　　　4. 广义相对论

5. 狭义相对论　　　　　　　　6. 牛顿力学

7. 大爆炸理论　　　　　　　　8. 原子弹

9. 耗资巨大的科学研究　　　　10. 质量守恒定律

Unit Eight　Modern Science and Technology (I)

11. 元素周期表　　　　　12. 有机化学
13. 生理化学　　　　　　14. 键角
15. 氨基酸类　　　　　　16. 产褥热
17. 生源说　　　　　　　18. 预防医学
19. 狂犬病疫苗　　　　　20. 自然选择
21. 遗传工程　　　　　　22. 人类基因组计划

II. Translate the following sentences into Chinese.

1. The behavior of electricity and magnetism was studied by Faraday, Ohm, and others during the early 19th century. These studies led to the unification of the two phenomena into a single theory of electromagnetism, by Maxwell (known as Maxwell's equations).

2. Currently, general relativity and quantum mechanics are inconsistent with each other, and efforts are underway to unify the two.

3. The synthesis of urea by Friedrich Wöhler opened a new research field, organic chemistry, and by the end of the 19th century, scientists were able to synthesize hundreds of organic compounds. The later part of the 19th century saw the exploitation of the Earth's petrochemicals, after the exhaustion of the oil supply from whaling.

4. Darwin proposed that the features of all living things, including humans, were shaped by natural processes over long periods of time. Implications of evolution on fields outside of pure science have led to both opposition and support from different parts of society, and profoundly influenced the popular understanding of "man's place in the universe".

5. Mendel's laws provided the beginnings of the study of genetics, which became a major field of research for both scientific and industrial research. By 1953, James D. Watson, Francis Crick and Maurice Wilkins clarified the basic structure of DNA, the genetic material for expressing life in all its forms.

Unit Nine
Modern Science and Technology (II)

Part A Lecture

猜测词义

我们在阅读过程中常常会碰到一些不认识的词语。如果这些词语不影响对文章主要内容的理解，我们可以略过它们，继续阅读；但如果它们成为理解文章重要内容的障碍，我们就必须设法弄清它们的意义。弄清词义最简单的做法是查字典，但一遇到生词就查字典会减慢阅读速度，打断阅读思路，影响阅读理解。因此，为了提高阅读效率，我们需要判断什么情况下必须弄清生词的意思；什么情况下遇到生词可以置之不理，继续阅读。这就需要运用对付生词的技巧，即猜测词义的各种能力。猜测词义是一项非常有用的阅读技能，这项技能的获得有助于正确理解阅读材料，提高阅读效率。本单元的讲座将介绍利用上下文线索和构词法猜测词义的技巧。

1 利用上下文线索和构词法猜测词义的技巧

生词是外语阅读中经常遇到的问题，是阅读过程中的一个障碍。要提高阅读的速度和阅读理解能力，在阅读过程中遇到生词时，最好不查字典而是依靠上下文提供的线索或进行构词分析来猜测词义。猜测词义是指在阅读过程中，从对语篇中的信息、逻辑、背景知识及语言结构等的综合理解出发，去猜测某些生词的词义。下面我们将分别介绍一些最常用的利用上下文线索和构词法来猜测词义的技巧和方法。

1·1 利用上下文线索猜测词义

段落或篇章是一个语义整体。一个词或词组在文中不是孤立存在。它的意义和句中其他词语的意义，以及该句前后句乃至整段或整篇的语境（上下文）

Unit Nine Modern Science and Technology (II)

有密切的联系。上下文的含义很广，大的方面可指语域和背景知识，小的方面则指生词前后的词语和句子。在科技英语文献阅读中，语域和背景知识对猜测和理解生词词义是非常重要的。因此，在阅读理解时，我们应充分利用上下文的有机联系，对生词的词义进行猜测。试分析下例。

Precipitation, commonly referred to as rainfall, is a measure of the quantity of water in the form of either rain, hail or snow which reaches the ground. The average annual precipitation over the whole of the United States is thirty-six inches. It should be understood however, that a foot of snow is not equal to a foot of precipitation. A general formula for computing the precipitation of snowfall is that thirty-eight inches of snow is equal to one inch of precipitation. In New York State, for example, seventy-six inches of snow in one year would be recorded as only two inches of precipitation. Forty inches of rain would be recorded as forty inches of precipitation. The total annual precipitation would be recorded as forty-two inches.

(1) The term "precipitation" includes _____.
   A) only rain   B) rain, hail, and snow
   C) rain, snow, and humidity   D) rain, hail, and humidity
(2) Another word which is often used in place of precipitation is _____.
   A) humidity   B) rainfall
   C) wetness   D) rain-snow

分析：短文的第 1 句对"precipitation"进行了定义与解释。"precipitation"是用来计量以雨、冰雹或雪的形式降到地面的水量的。它常常被称为"rainfall"。此外，文中接下来对纽约州的"annual precipitation"的计算，对我们的推测也有一定的帮助。

以上例子说明，猜测词义需要一定的技巧。常见的技巧包括以定义、重述、同义和反义关系、举例、经验和常识、相关信息等为线索进行词义猜测。

▶ 1.1.1 以定义为线索猜测词义

为了帮助读者理解文章的内容和实质，作者往往会对那些读者不熟悉的词汇、术语、概念下定义，以简练的语言说明这些词语在文中的意义。在科技英语文献中，定义的使用尤为频繁。如果读者能识别定义线索，对猜测词义和概念是大有裨益的。定义线索主要是通过定义句的标记动词、同位语或定语从句、标点符号等给出定义性暗示。

（1）以定义句的标记动词为线索

定义句通常用标记动词 be、mean、define、call、know、term 和 refer to 等引导。如：

A movement of the head, hand or body to express a feeling is called a **gesture**.

分析：根据动词后给出的定义，可以很容易地猜出 gesture 为"手势，姿势"。

（2）以同位语或定语从句为线索

许多定义性暗示是通过同位语给出的。如：

Ethology, the scientific study of animal behavior, is one of such disciplines.

分析：同位语 the scientific study of animal behavior 提示了 ethology 的词义，由此推出其为"动物行为学"之意。

除同位语外，有些限定性定语从句和非限定性定语从句也起到解释说明作用，可以作为猜测词义的线索。如：

The Lincoln Memorial in Washington D. C. is not unlike the **temples** that ancient Greeks built in honor of their gods.

Unit Nine　Modern Science and Technology (II)

分析：句中 that 引出的定语从句以定义性的叙述提示了 temples 是为祭神而修建的建筑物，即"寺庙，圣殿"。

（3）以标点符号为线索

简短的定义可由逗号、冒号、破折号、圆括号、方括号等引出。这些标点符号担当提示信号，给出定义性暗示。如：

Fibrin, elastic threads of protein, helps blood to clot.

分析：句中的 fibrin 一词用两个逗号之间的部分加以解释说明，意思是"纤维素"。

综上所述，许多词或短语的定义性暗示在文中出现前都可能有信号标志。这些解释性词汇与标点符号一起为理解生词提供了线索。我们应当敏锐地识别这些信号，积极地利用信号后出现的暗示来猜测词义。

▶ 1.1.2 利用重述猜测词义

重述是作者使用读者熟悉的词语，对已出现的陌生词语进行的解释。它所给的暗示虽不如定义直接明了，但也能为我们推测词义提供足够的信息。为了提醒读者下文将出现的暗示，作者往往借助插入语为信号。有关表示重述的信号词有 or、or rather、that is、in other words 和 namely 等，如：

Scientists believed that there was a deep **crevasse**, that is, a deep open crack, under the ice through which a submarine could pass.

分析：信号词 that is 提示下文将出现对 crevasse 的词义解释，由此不难猜出 crevasse 的词义为"缺口，裂缝"。

重述暗示有时虽无信号词语可寻，但从作者的重复叙述中，或从上下文中的逻辑关系和语义联系中可以找到一些猜测生词的线索。如：

251

Carbon monoxide (CO) is a **noxious** gas which can cause death.

分析：从定语从句的重述中可以猜出此句中 noxious 的词义是"有毒的"。

1.1.3 利用文章中词与词的同义和反义关系猜测词义

作者有时将生词与熟悉的词进行比较，指出它们的共同点或不同点，以帮助读者理解。我们可以充分利用这种线索来猜测词义。

（1）以同义词、近义词为线索猜测词义

英语中同义词、近义词特别丰富，作者有时把两个或更多个同义词或意义相近的词用在一起，使自己的意思表述更明确，或在修辞上达到强调或避免单一乏味的效果。如：

Many famous psychologists are trying to understand the problems modern people suffer from, but even these **eminent** scholars are confused about what causes them.

分析：句中 eminent 与 famous 同义，意为"著名的"。

同义关系可以通过信号词（or、like、as...as、the same as 等）、标点符号和不同词的使用来体现。如：

Their friends laughed at them, and my sister felt **wretched**, very unhappy.

分析：句中"wretched"与"unhappy"意思相近，意为"不幸的，可怜的"。

Doctors believe that smoking cigarettes is **detrimental** to your health. They also regard drinking as harmful.

分析：第 2 句中的 they、also、drinking 和 harmful 与第 1 句中的生词 detri-

252

Unit Nine Modern Science and Technology (II)

mental 发生联系，帮助我们猜出 detrimental 的词义是"有害的"。

（2）以反义词和对照关系为线索猜测词义

为使表述的意思更明确，或在修辞上收到强调的效果，有时作者会把意义相反的词成对地放在一起。如：

Questionnaires have certain obvious advantages, but they also have **drawbacks. Spontaneous** answers cannot be distinguished from thought-out answers. Questions can be misunderstood because it is difficult to avoid ambiguity except in the most simple questions. Different answers cannot be treated as independent since the subject can see all the questions before answering any one of them.

分析：从第 1 句可知，advantages 与 drawbacks 形成对照，由此我们推断 drawbacks 的意思是 disadvantages，表示对照的信号词是 but。从第 2 句可以推测 a spontaneous answer（即兴回答）不是 a thought-out answer。换句话说，spontaneous 的意思应该是 without thinking（不假思索地）。

反义关系可以通过信号词（not、but、while、however、unlike、on the other hand、in contrast to 等）和意义相反的词的使用来体现。如：

In the northern regions the winters are generally cold and **humid**, and the summers hot and dry.

分析：显然，冬天和夏天的气候是截然相反的，它们的修饰词的意思也应该截然相反。"cold"与"hot"对应，"humid"与"dry"对应。因此，"humid"是"潮湿"的意思。

▶ 1.1.4 利用举例为线索猜测词义

用实例来解释和说明生词是一种常见的现象。作者利用词与词之间的上下义关系给读者以暗示。举例通常是给出下义词（或局部性词）以提示读者猜测上义词（或概括词）的词义，也就是从具体事例中归纳出一般概念。举例常常提供有价值的暗示,可以帮助我们猜测词义。常见的提示举例的信号是冒号以及 for example、for instance、such as 等。如:

Today young couples who are just starting their households often spend lots of their money on **appliances**, for instance, washing machines, refrigerators and color TVs.

分析:通过所举的例子(washing machines, refrigerators and color TVs)可以看出,"appliances"应是这些名词的总称,即"家用电器"。

应该注意的是,有时作者在举例暗示前并不给读者发出任何信号。这时,我们可以根据语义,判断上下义关系、局部与整体的关系,从而猜测词义。如:

She is studying **glaucoma** and other diseases of the eye.

分析:我们虽不能猜出 glaucoma 的确切意义,但可以从概括词"眼病"以及 other 的提示中,推断出 glaucoma 是下义词,指某种眼病。

▶ 1.1.5 根据经验和常识猜测词义

在阅读过程中,一个人的经历和阅历(即经验和常识)对猜测词义和理解文章内容起着十分重要的作用。经历越丰富、知识面越广的人,其猜测生词的能力就越强。如:

A pair of **spectacles** is a device for correcting eye-sight. It consists of three main parts: a lens for each eye and a frame.

分析:凭我们的常识可以猜出 spectacles 的意思是"眼镜"。

▶ 1.1.6 根据相关信息综合猜测词义

除了上述技巧外,从上下文中发现并利用与某一概念相关的信息,并把

Unit Nine　Modern Science and Technology (II)

信息综合起来考虑，通过合理概括，同样能为我们推测有关词语的意义提供足够的线索。如：

The fishermen make their **canoes** from tree trunks. They go from island to island in these light (weight) marrow (shape) boats and collect turtles' eggs.

分析：我们从上下文中可以得出以下信息："canoes"是一种渔夫用树木做的、来回于岛屿之间的、重量轻且形状狭长的、类似于小船之类的东西。尽管我们可能还不能肯定它的确切解释，但这一生词已经不会影响我们的阅读和理解。

1·2 利用构词法猜测词义

分析和了解英语单词的结构是在阅读中进行词义猜测的另一种方法。国内外研究外语阅读教学的专家学者们一致指出，想要培养学生的阅读能力和速度，就要着力培养学生的构词分析能力，通过词素来辨认、理解那些不认识的词汇，也就是通过常用的词根和词缀去分析新词的含义，去认识同族词（同根词）中许多原来不认识的新词，在上下文中去猜测词义。如：

A. They **overestimate** the interviewee's ability and asked him many difficult questions.

分析："estimate"是"估计"的意思，"over-"是前缀，意为"过分，过度，超过"等。因此，"overestimate"就是"高估"的意思。

B. We were told that ours was the most **spacious** room in the hotel. That was why we had to pay so much for it.

分析："spacious"是由"space（名词，空间）"+"-ious（形容词后缀）"变化而来的。因此，可猜测其词义为"宽敞的"。

我们知道，英语单词是由词根和词缀两大部分组成的。而词缀又分为前缀和后缀。语言学家把词根、前缀和后缀称为"扩大词汇量的三把钥匙"。如果较好地掌握了构词的"三把钥匙"，词汇运用能力就会得到很大提高，在阅

255

读科技英语文献时更会从中受益匪浅，因为许多英语科技词汇都是从希腊语和拉丁语的词根派生出来的，而且相当数量的科技领域里的新词和术语都遵循构词法。因此，我们对词的结构进行准确的分析，也能"望形生义"。通过构词法猜测词义，通常从词的构成部件——词根和词缀下手。

（1）词根（root）

词根是词的核心意义所在。推测词义，如能从词根出发，再结合词缀和上下文，可达到事半功倍的效果。如：

Mary **collaborated** on a novel with her friend.

A) worked together B) read together

C) jointly bought D) jointly supported

分析：我们如果知道句中的斜体词的词根 labor- 意为 work 的话，即可确定答案为 A。

从上例可知，词根是猜测词义的基础。实际上，词根是具有构词能力的基本词义单位。在同一词根上加不同的词缀可以表示不同的意思。我们在英语学习过程中要特别注意构词能力最强的词根。

（2）词缀（affix）

词缀分为前缀和后缀。前缀和后缀能增加或改变词根的意义，如：

(1) Unlike decisions which are made on the basis of mutual concerns, **unilateral** decisions can be unpopular because they are made by only one of the parties concerned.

A) one-sided B) two-sided C) many-sided D) equal-sided

(2) She has a wide interest and is an **omnivorous** reader.

A) all-reading B) critical C) fast D) efficient

分析：句 (1) 中，词根 -later 意为 side，前缀 uni- 意为 one，故答案为 A；句 (2) 中，词根 -vor- 表示 eat，前缀 omni- 表示 all，这里由 all-eating 引申为 all-reading，故答案为 A。

Unit Nine Modern Science and Technology (II)

前缀一般只引起意思上的变化,而不构成词类的转变。它们可以表示方式、态度、程度、时间、位置等概念。和前缀不同,后缀附加在词或词根的后面,改变词类,对词义的影响不明显。后缀分名词后缀、动词后缀、形容词后缀和副词后缀。了解常用前缀和后缀可以扩大词汇量。

综上所述,各种各样的前后缀可以构成名词、形容词、动词、副词等,但这些词缀需要平时不断地积累和记忆。掌握构词法知识是扩大词汇量和猜测生词词义的最佳办法。

2 综合练习

上节我们介绍了猜测词义的一些常用技巧,本节将综合运用上面所讲的技巧进行词义猜测训练。值得提醒的是,对某一特定词词义的推测,通常不是只靠某一种技巧,而是借助多种技巧的"共同合作",即采用综合性猜词策略。综合性猜词策略是把与生词有关的信息进行综合考虑、推断和检验的过程。它需要运用包括语法结构、词性、上下文的各种关系及构词法等在内的多种知识和线索。这种策略的具体步骤如下:1)观察生词本身的特点及在句中所处的位置,确定该词的词性;2)看一看生词所处的上下文,即生词所处的分句或整个句子,确定该词与其他词之间的关系;3)看一看生词所处的句子与其他句子或段落之间的逻辑联系,找出信号词、标点符号或参照词等线索;4)利用所获得的知识来推断词义;5)检查推断是否正确。

Although many of them say they will give their support, their actions **belie** their words.

分析:1)从生词本身及其在句中所起作用来看,belie 应是一个动词。2)从生词所处的句子来看,belie 的主语是 their actions,宾语是 their words。3)Although 作为信号词表明了 belie 所处的主句与 although 所处的从句之间是一种让步关系,由此看出 belie 所在的主句含有与从句相反的信息。4)在前面三步的基础上,可以对 belie 的词义进行推断,belie 与 contradict 的意思接近。5)检查这个推断是否正确。contradict 也是动词,意为"相矛盾;不一致"。代入句中"their actions contradict their words",

即：它们的行动与它们所说的相矛盾。意思合乎逻辑。而 belie 在词典中的意思为"与……不一致/不相等"。

Wine is the fermented juice of fresh grapes. The juice of the wine grape contains sugar, and growths of **yeast** form on the outside of the grape skins. In wine-making, the grapes are crushed in a wine press and the **yeast** converts the sugar to alcohol, when there is no air present, by a process called fermentation. Red wine is made from dark grapes, and white wine from white grapes or from dark grapes whose skins have been removed from the wine press at an early stage.

Beer is made from sprouting barley grains (malt) which is fermented with **yeast** to produce alcohol; hops are added for flavor. Ale, the most common drink in England in the Middle Ages, was also made from barley, but without hops; the ale of today is merely a type of beer. In Japan beer is made from rice.

分析：从 yeast 的形式和它在句中的位置，可以知道它是一个名词。从短文中我们知道它生长在葡萄皮的外面，因此它大概是一种小的植物；我们还知道它可以将糖转换成酒精，能将葡萄汁变成葡萄酒，还能将大麦（barley）变成啤酒。由此，我们已经有了足够的信息可以进行推断，yeast 很可能是指"酵母"。

综上所述，在阅读过程中，我们要具有即使不了解生词词义也不影响理解的能力；在猜测词义时，我们要充分利用语法结构和标点符号方面的线索，确定该生词与句子其他部分的关系，并利用句中（段中）其他词的意思和整个句子的意思来缩小词义的范围。同时，在对具体的某一词义的推测过程中，我们须综合利用上下文所提供的意义线索、构词法知识和自己的经验和常识，才能取得好的效果。

Unit Nine　Modern Science and Technology (II)

Read the following paragraphs, and choose the best answer to each question.

1

Kids operating computers **implement** their curriculum with great versatility. A music student can program musical notes so that the computer will play Beethoven or the Beatles. For a biology class, the computer can produce a picture of the **intricate** actions of the body's organs, thus enabling today's students to envisage human biology in a profound way. A nuclear reactor is no longer an **enigma** to students who can see its workings in **minute** detail on a computer. In Wisconsin, the Chippewa Indians are studying their ancient and almost forgotten language with the aid of a computer. More commonly, the computer is used for drilling math and language concepts so that youngsters may learn at their own speed without trying the patience of their human teachers. The simplest computers aid the handicapped, who learn more rapidly from the computer than from humans. Once **irksome**, remedial drills and exercises now on computer are conductive to learning because the machine responds to correct answers with praise and to incorrect answers with frowns and even an occasional tear.

(1) The word "implement" in Sentence 1 means _____.
　　A) learn　　　　　　　B) add to
　　C) make use of　　　　D) plan

(2) In Sentence 3, the word "intricate" can be replaced by _____.
　　A) complex　　　　　　B) puzzling
　　C) detailed　　　　　　D) interior

(3) The word "enigma" in Sentence 4 refers to _____.

    A) energy   B) problem

    C) mystery   D) trial

(4) The word "minute" in the same sentence means _____.

    A) 60 seconds   B) tiny

    C) multiple   D) hidden

(5) In the last sentence, the word "irksome" is closest in meaning to _____.

    A) easy   B) pleasant

    C) sick   D) boring

2

[1] The Bermuda Triangle is an area of the Atlantic Ocean which has puzzled people for several centuries.

[2] There are a number of environmental features that can be found only in the Bermuda Triangle. Because of these unique characteristics, the triangle is one of the most dangerous areas in the Atlantic Ocean.

[3] The weather in the triangle is **treacherous**; its sudden changes often endanger the lives of sailors.

[4] The triangle is well known for unexpected storms, hurricanes that are out of season, and other unnatural events. Many people feel that this freak weather can explain most of the strange events which have occurred there.

[5] Because of the many violent storms, the triangle is often **impassable**.

[6] Many ships leave land and disappear completely; the U.S.S. Cyclops, for example, **vanished** in 1918.

Unit Nine　Modern Science and Technology (II)

[7] Some missing ships carried cargo such as coal, oil, mahogany, and military supplies, while other ships carried only passengers.

[8] Engine **malfunction** might explain the disappearance of old ships, but investigators must look for other explanations when a new ship disappears.

[9] When a ship is reported missing, searchers rush to the area to look for evidence which might explain the disappearance.

[10] One author tries to explain the disappearances of the ships by attributing them to natural events or human error.

[11] Another author believes that intelligences from another world are responsible for the disappearances. He says that these **extraterrestrial beings** have a zoo where they kept all the missing seamen.

[12] Many people agree that natural causes cannot explain the strange events in the triangle. "It just isn't natural for a ship to completely disappear like this. It's weird; no matter how you try to explain it," declared the seamen.

[13] "These **eerie** events make me afraid to sail out of sight of land."

(1) The word "treacherous" in Paragraph 3 is treacherous means _____.

    A) powerful   B) changeable

    C) different   D) tremendous

(2) The word "impassable" in Paragraph 5 most probably means _____.

    A) impracticable   B) not practical

    C) not conquerable   D) blocked

(3) The word "vanished" in Paragraph 6 may be substituted by _____.

    A) got lost   B) got stuck

    C) faded away   D) passed away

(4) The phrase "extraterrestrial beings" in Paragraph 11 most probably refers to _____.

    A) extremely intelligent zoo keepers

    B) underwater animals

    C) extra-ordinary beings

    D) creatures from out space

(5) The word "eerie" in the last sentence may be a synonymy of _____.

    A) weird   B) strange

    C) unnatural   D) all of the above

(6) The word "malfunction" in Paragraph 8 is closest in meaning to _____.

    A) failure   B) mistake

    C) horse power   D) agedness

Part B Reading

Chips and the Internet[1]

As sociologists Hutchby and Ellis (2001) put it, "information and communication technologies (ICT) are set to wreak widespread social, cultural economic and political change in the twenty-first century". Consequently, computers in particular came to assume status symbol much in the same manner television sets or even washing machines had four decades earlier. This passage tells how computer revolution changed human life more drastically than all that had preceded.

Computers represent another domain where applied science has transformed technology and society. The historical roots of the modern electronic computer go back to the automated Jacquard loom (1801), which used punch cards to control weaving patterns. Another stream in this river of technological change flowed from the calculating needs of industrializing society and notably the mechanical "calculating engines" of Charles Babbage (1791–1871) who conceived of a universal calculating machine or computer. Improved mechanical and electro-mechanical calculators followed through the 1930s; these were underwritten in part by the U.S. Census Bureau and used for producing numerical tables and processing data. Not unexpectedly, World War II pushed developments in computing that resulted in the Mark I calculator in 1944 and the Navy-sponsored ENIAC in 1946. After the war, a series of ever more sophisticated general-purpose, stored-program digital electronic computers emerged, including the UNIVAC machine, the first commercially available and mass-produced central computer. It appeared on the market in 1951. This type of general-purpose computer run by programs was based on principles articulated in 1937 by Alan Turing (1912–1954) and amplified by Turing and by John von Neumann, among others, through the 1950s.

The key scientifico-technical innovation for the development of computers was the creation of the first solid-state transistor in 1947 by William Shockley, a mechanical engineer, and his team of scientists working at Bell Laboratories, an achievement for which he won the Nobel Prize in physics in 1956. Such solid-state devices supplanted vacuum tubes (which regularly failed), and they permitted the industrial and commercial development of the first practical large computers in the 1950s and 1960s, such as the early standard, the IBM 360. The miniaturization of the transistor and the creation of the semiconductor computer chip was a related triumph by materials scientists and engineers. The exponential growth of the computer revolution was under way, a pattern of growth captured by Gordon Moore's "law", enunciated in 1965: It states that the number of electronic switches on semiconductors doubles every eighteen months, a technical way of expressing the (surprisingly accurate and meaningful) concept that, in our era at least, the computer chips produced by industry are always much more powerful than the equivalent chips only a year or two before. This prediction still holds true four decades later as physicists explore computing possibilities on the nanometer (10^{-9} meter) level, although computer scientists speculate that we must be approaching the "repeal" of Moore's Law sometime in the next generation, as the paths of transistors approach the width of an atom. From the 1960s onward mainframe computers manufactured by the IBM Corporation, Digital Equipment Corporation, and other companies came into increasingly widespread use in the military, in the banking and emerging credit card industries, in airline reservation systems, in and around stock markets, and on university campuses wired with terminals hooked up to central multiprocessing and timesharing machines and printers. The computer center became a new social space.

In a host of realms physical scientists, computer scientists, and computer engineers toiled to improve the amazing technology of the computer: software and programming languages; the silicon "chip" itself and ever more elaborate and faster integrated circuits; high-tech methods to manufacture these chips; increasingly sophisticated magnetic media and data storage technologies (the "floppy disk"); improved means of displaying information; and industry

Unit Nine   Modern Science and Technology (II)

standards in general. But nothing did more to unleash the computer and information revolution than the introduction of the personal computer (PC) in the 1970s and 1980s. Hobbyists began to build personal computers from kits and commercially available microprocessors and floppy disks in the early 1970s. In 1977 college dropouts Steve Jobs and Steve Wozniak brought to market the Apple II, and in 1981 IBM branded the name PC with its own machine powered by the DOS operating system licensed from Microsoft Corporation, run by a young Harvard dropout named Bill Gates. Refinements in the field of personal computing included the mouse and the graphical user interface with its folders, windows, and icons, developed at the Palo Alto Research Center of the Xerox Corporation. Improved operating systems and ever more sophisticated software packages for word processing, spreadsheets, and databases transformed the new toys into indispensable machines for the office, the home, and modern life in general. The computer became a mighty and expensive consumer item, and businesses slow to invest in the technology paid a price. In 1981 the Commodore company's VIC-20 alone sold one million units. In 1989 Microsoft revenues reached one billion dollars. A radical transformation was under way.

Nobody reading this needs to be informed of the explosive proliferation of computing devices. Computers are ubiquitous, and they exist at the core of industrial civilization to the point that industrial civilization as we know it would be impossible without them. Essentially no one anticipated the computer, however, or how it would transform society. Yet computing and personal computing have penetrated deeply into economies and societies, and in barely two decades the computer revolution has effected an enormous social transformation. The number of personal computers and computing systems of all sorts is substantial and growing exponentially. But the computer revolution is a revolution in progress. In the year 2000, for example, the most intensely computerized country was the United States, with 161 million personal computers. Although this is a significant number, it represents only slightly more than one computer for every two Americans. On a world basis, in 2000 there were only 450 million PCs or one computer for roughly every seven persons, a figure suggesting that the social penetration of the computer has considerably further to go.

The extraordinary novelty and perhaps the key to understanding the computer as technological agent of social change was the idea of connecting computers together into networks. First it became possible to connect a remote terminal with mainframe computers over phone lines using modems. Then, funded by DARPA (the Defense Advanced Research Projects Agency), the first true computer network, ARPANET, arose around a handful of nodes in 1969. ARPANET grew to link more and more universities and research labs, and the utility of the network began to demonstrate itself not least in the new modality of e-mail. By the 1990s the Internet (and later the World Wide Web) stood poised for commercial development, and in the last two decades the world has witnessed the development of the incredible and transformative global information system that the Internet is today.

The Internet is only partly about the routers or servers that make up the physical system, and it is only partly about the Internet Service Providers (ISPs) and the economics of Web-based sales or services. The Internet is mostly about communication and access to information. The Internet serves as the substrate and technical basis for personal communication through e-mail. People around the globe now have instant access to one another via e-mail and the Internet. But no less significant, in what amounts to being a universal library, the Internet also brings the world's information instantaneously to video monitors in homes and offices and on the road everywhere across the globe. Commercial search engines comb the Web and fulfill every request for information and place it immediately at our disposal. The ancient Library at Alexandria is less than nothing compared to the libraries of information available on demand over the global Internet today. Computers and the Internet now give us information of unimaginable variety at the click of a mouse. Some of this information is idiosyncratic and unreliable, but much of it is credible and authoritative; some of it may be morally or otherwise offensive, but it is all there. The Internet has also emerged as a new global marketplace with virtually unlimited retailing, including familiar companies like the

Unit Nine   Modern Science and Technology (II)

booksellers Amazon.com. Other sites, such as eBay, have become familiar landmarks in the netherworld available to us. While many caveats need to be introduced, the effect of the computer revolution and the Internet has been to liberate and democratize information. Those with access to computers have access to information and can make their own judgments; they are less in the grip of traditional experts and guardians of information. Those without computers and access to this Internet, of course, are denied this capability. The powerfully global and globalizing technology of the Internet has the paradoxical effect, not of dominating people by limiting their choices, but of allowing individuals the freedom to exploit the Internet according to their own priorities and less according to any "establishment" or political authority. The Internet fosters a sense of community among users who share common interests to the point that cyberspace may be thought of as the jostling of splintered groups sharing shifting interests. The Internet is more interactive and more intimate than television, a more passive medium long derided for its remote control, which often turns its users into proverbial couch potatoes.

Computers and digital recordings have completely transformed the music industry, which is now based on computer processing and digital storage and retrieval, notably through laser-read and -written disks. A word is necessary about computer and video games. With ravenous appetites for speed, processing power, and storage, computer gaming has been a major driver of the computer revolution ever since the introduction of the primitive game PONG by the Atari company in 1976. Engrained in youth subcultures around the world, computer games have become exquisitely complex, with ever-improving capabilities for graphical representation. Today, it costs fifty million dollars to develop and introduce a new retail computer game, and a successful game is worth many millions more. The stakes for entire gaming systems are even higher. Along these lines, developments initially for the military of supercomputers and supercomputing allowed for improved handling of graphical and visual representations of information. Today, larger and faster personal computers are capable of handling some of these intensive computational tasks, including games, digital photography, and digital home video. Assemblages of supercharged PCs can handle the processing needs met by the supercomputers of just a few years ago, and these machines provide the computing basis for related developments in artificial intelligence and data mining. Finally in this connection, computers and computing power have crossed over to the movie industry. Digital projection systems are now common, and an entire subgenre of digital animation and digitally enhanced movies has arisen. Today, many films incorporate aspects of digital and computer-based images that were impossible prior to the advent of computers. Developments regarding the computer are recent, practically too current to be called "history", and they testify yet again to the place of high-tech industries and the rapid pace of change in industrialized civilization in the twenty-first century.

New Words and Expressions

underwrite 担保，承保
supplant 取代，替代
exponential 快速增长的；指数的
enunciate 清楚地表明，阐明
speculate 推测，猜测
repeal 废除，撤销，废止
toil 苦干，辛勤劳作
unleash 发挥；放开
kits 工具箱
dropout 辍学者，退学者
icon 图标；崇拜对象，偶像
revenue 收入，收益
proliferation 激增；涌现；增殖
ubiquitous 无处不在的
modem 调制解调器
node 节点

poised 处于准备状态；蓄势待发
substrate 基底；基层
comb 梳理；搜寻
idiosyncratic 乖僻的；怪异的；特殊的
caveat 警告，告诫
grip 紧握；（对……的）控制，影响力；理解
paradoxical 似自相矛盾的
cyberspace 网络空间
jostle 推挤
splinter 裂成碎片，分裂
deride 嘲笑，愚弄
ravenous 贪婪的；极饿的
engrain 使根深蒂固
subculture 亚文化行为观念，次文化

Unit Nine Modern Science and Technology (II)

> exquisitely 精巧地；敏锐地
> stake 赌注；股份
> assemblage 集聚
> supercharged 格外强劲的
> subgenre 次类型；亚体裁
> testify 证实，证明

Notes

1. 本文节选自 James E. McClellan III & Harold Dorn 的 *Science and Technology in World History: An Introduction* 一书。

Exercises

I. Find out the English equivalents to the following Chinese terms from the passage.

1. 自动提花织机
2. 计算引擎
3. 数值表
4. 固态晶体管
5. 真空管
6. 摩尔定律
7. 中央多处理器
8. 分时机
9. 磁体媒介
10. 数据存储技术
11. 软盘
12. 个人计算机
13. DOS 操作系统
14. 图形用户界面
15. 电子表格
16. 万维网
17. 服务器
18. 因特网服务提供商
19. 社区意识
20. 家庭数字化视频
21. 数据挖掘
22. 数字放映系统

II. Translate the following sentences into Chinese.

1. Not unexpectedly, World War II pushed developments in computing that resulted in the Mark I calculator in 1944 and the Navy-sponsored ENIAC in 1946. After the war, a series of ever more sophisticated general-purpose, stored-program digital electronic computers emerged, including

the UNIVAC machine, the first commercially available and mass-produced central computer.

2. The miniaturization of the transistor and the creation of the semiconductor computer chip was a related triumph by materials scientists and engineers.

3. From the 1960s onward mainframe computers manufactured by the IBM Corporation, Digital Equipment Corporation, and other companies came into increasingly widespread use in the military, in the banking and emerging credit card industries, in airline reservation systems, in and around stock markets, and on university campuses wired with terminals hooked up to central multiprocessing and timesharing machines and printers.

4. But nothing did more to unleash the computer and information revolution than the introduction of the personal computer (PC) in the 1970s and 1980s.

5. Computers are ubiquitous, and they exist at the core of industrial civilization to the point that industrial civilization as we know it would be impossible without them. Essentially no one anticipated the computer, however, or how it would transform society. Yet computing and personal computing have penetrated deeply into economies and societies, and in barely two decades the computer revolution has effected an enormous social transformation.

6. By the 1990s the Internet (and later the World Wide Web) stood poised for commercial development, and in the last two decades the world has witnessed the development of the incredible and transformative global information system that the Internet is today.

7. The ancient Library at Alexandria is less than nothing compared to the libraries of information available on demand over the global Internet today. Computers and the Internet now give us information of unimaginable variety at the click of a mouse.

8. Developments regarding the computer are recent, practically too current to be called "history", and they testify yet again to the place of high-tech industries and the rapid pace of change in industrialized civilization in the twenty-first century.

Part C Extended Reading

In 1900, They Never Imagined

Technology is rapidly changing our world. It is bringing us services beyond our grandparents' wildest dreams. This passage describes major changes in scientific discovery and technological progress in the past 100 years.

The computer revolution

In 1900, they never imagined…

1944 The Harvard-IBM Mark I [1] computer

1946 The 24-meter long ENIAC [2] computer

1964 The IBM System 360 mainframe [3]

1965 The Digital PDP [4] -8 mini-computer

1974 The Altair 8800 mini-computer and the communications system ARPANET [5]

1981 The IBM Personal Computer [6]

In 1900, they never imagined…

The computer and its miniaturization. Thanks to more and more capacity for memory processing through smaller and smaller silicon chips, the mainframe computer which needed a large room became a desktop computer and workstation, which eventually even had greater capacity.

The personal computer, revolutionized by the introduction of IBM's Personal Computer (PC) in 1981, spread to 245 million PC users by the century's end. The PC was not only word processor, business organizer, research and educational tool, home study center and games player, but allowed global communication through the Internet.

The Internet

The Internet was prefigured by the communications system ARPANET, initially a US Defense Department network expanded by universities. The Internet is a linked computer communications network used for information, e-mail, business and education. US-based International Data Communications predicts the Internet economy will reach $ US 1 trillion by 2002.

In 1900, they never imagined...

1977 Diagnosis of a car's mechanical problems by laptop computer.

1998 The wrist watch becoming a miniature computer.

1996 The notebook in the form of a portable laptop computer.

The Man/Woman of the Year displaced by a machine, in *Time* magazine's prophetic recognition of the dominant future of the Personal Computer. Instead of Man of the Year or Woman of the Year, in 1982 *Time* magazine nominated the computer as Machine of the Year.

The air transport revolution

In 1900, they never imagined ...

That there would be flight by heavier-than-air machines, beginning with Orville and Wilbur Wright's [7] Flyer in 1903, reaching a climax in 1969 with the first flight of the Boeing 747 Jumbo jet which can carry 400 people.

In 1900, they never imagined ...

Space travel. The space shuttle, able to ferry men and women to and from an orbiting space station, was put into service from 1981, 20 years after the first man in space, Yuri Gagarin [8].

Mass air transport. This was begun with the development of small passenger planes after World War I. At the century's end, the top 20 airlines carry more than 500 million passengers annually.

Supersonic flight. The Concorde [9] in commercial service from 1976, cruises at twice the speed of sound, and makes the trans-Atlantic flight in 3 1/2 hours, compared to seven hours for a Jumbo jet.

Unit Nine   Modern Science and Technology (II)

The telephone revolution

In 1900, they never imagined ...

That people would be able to walk through the street carrying on a conversation with someone in distant cities by using a tiny, hand-held mobile phone. They never envisaged that messages could be sent and received from such a machine through the use of email, fax and electronic paging.

In 1900, they never imagined ...

1901 Telegraph poles carrying up to 150 telephone wires each.

1919 The first Dial Telephone in general use.

1959 The first mobile in-car phone by Bell Telephone, US.

1973 The first portable cellular phone by Motorola, US.

1978 The first city cellular radio mobile radiotelephone system, Chicago, US. Systems linked to new phone communication technology included 1991 Cordless Phones.

1993 Pagers, alerting wearers to phone messages.

1996 Teleconferences, using a speaker-locating camera for video images.

1996 The portable fax machine, hooked up to a portable phone.

The science revolution

In 1900, they never imagined ...

That Max Planck's [10] quantum theory, introduced in 1900, and which changed our understanding of the physical universe, would give rise to ...

1905 Albert Einstein's [11] Special Theory of Relativity, and in ...

1988 The research of Stephen Hawking [12], noted for his theories on black holes [13] in space. Hawking's best selling book on modern cosmology, *A Brief History of Time* [14] (1988) attempts to reunite quantum theory and Einstein's theory of relativity.

In 1900, they never imagined ...

Atomic research. In 1942, the first nuclear reactor was built by Enrico Fermi [15].

Medical research. Sir Alexander Fleming [16] discovered penicillin in 1928. Protection of wounded soldiers from blood poisoning sped up manufacture of the drug in World War II.

Organ transplants. The first kidney transplant operation was in 1954, the first heart transplant in 1967.

Genetic research. Genetic research escalated after the description of the DNA double helix in 1953. Following genetic experiment and breeding, the first living mammal was patented in 1988. In 1997 cloned sheep Dolly, in Roslin, Scotland, was created as a genetically identical sheep to her mother using DNA.

The disposable, consumable, convenience society

In 1900, they never imagined...

1901 The electric vacuum cleaner. Electric home appliances multiplied after the introduction of the electric vacuum cleaner (1901), washing machine (1907) and refrigerator (1910).

1907 The first heat-resistant plastic, Bakelite invented by Belgian, Leo Baekland [17], 1907. Plastic materials created this century included Teflon (Du Pont [18] 1937, trademarked 1945), Plexiglass (1937), polystyrene resins (1937), terylene (1941), polyethylene (1933, refined 1953) and polypropylene (1954). Ziegler [19] and Natta [20] won the Nobel Prize in 1963, for their polymer research in evolving polyethylene and polypropylene.

1924 Quick-frozen foods. Clarence Birdseye [21] discovered the benefits of quick-frozen food during a fur-trading expedition to the icy climate of Labrador from 1912 to 1916. Birdseye peas are one of the first frozen foods to be sold in small packets in the United States.

1930 Supermarkets. Benefiting from the frozen food revolution, supermarkets flourished after the success of the King Kullen self-service supermarket [22], New York in 1930.

1938 Nylon. Nylon, called the first high-performance engineering plastic, was created by Carothers [23] at Du Pont. When nylon stockings became

Unit Nine Modern Science and Technology (II)

generally commercially available in the US in 1940, four million pairs were sold in New York in one day.

1938 The ballpoint pen. The ballpoint pen with metal refills was invented by Hungarian Georg Biro [24] after he had devised viscous ink. Frenchman Marcel Bich [25] developed the plastic format in 1953.

1960 The contraceptive pill. The contraceptive pill, developed by Pincus [26] from 1954, was licensed for use in the Dr Gregory Us, and a year later in Britain.

1982/1983 The CD-ROM. The CD-ROM was invented by Philips and Sony Corporation. Laser technology and digital recording combine in an important recording and storage device.

The hi-tech revolution as commonplace

In 1900, they never imagined...

1906 The beginning of the public radio broadcasting, Massachusetts.

1936 The beginning of the first public television service, London. TV development sped up after the invention of the cathode ray tube "kinescope" by Vladimir Zworykin [27], head of RCA's TV research. This tube displaced John Logie Baird's [28] mechanical system. Worldwide television broadcasts are now possible through satellite technology.

1935 The first practical radar developed by Sir Robert Watson-Watt [29], England. Radar is extensively used today in air traffic control, tracking weather patterns and spacecraft.

1937 The Xerography electrostatic copying process. Chester Carlson [30] developed the Xerography electrostatic copying process, which was commercially produced in 1950. The first Xerox machine appeared on the market in 1958. Office photocopying is now worldwide. As machines become smarter, they can print on both sides, print color, sort in order and staple copies, and also act as printers instructed by computers.

1947 The transistor. The transistor was developed in the Bell Laboratories in the United States in 1947 after a long program of research. The transistor won for its inventors, John Bardeen [31], Walter H. Brattain [32] and William Shockley [33] the Nobel Prize for Physics in 1956 and was a key development in speeding up the way electric current was carried. The transistor was smaller, more robust, used less power and produced less heat than a valve.

The transistor began the process of miniaturization so important to late 20th-century electronics. The microchip was developed by Kilby [34] and Noyce [35] in 1959, the microprocessor by Hoff [36] at Intel in 1971. In 1988, Intel produced a 6cm-square microchip containing 5.5 million transistors.

1960 Laser technology, developed by Charles H. Townes [37] and Arthun Schawlow [38], Columbia University (1951), and Dr Theodore H. Maiman [39] working at the Hughes Research Labs [40] (1960). Laser technology is now vital in telecommunications, medicine, the defense industry, mass production, ozone measurements, and supermarkets (bar-code reading). Travelling through fiber optic cables, laser carry phone, fax, computer, TV and radio communications are used in medical operations, business, industry, commerce, construction and in the home.

1972 Calculators. Based on the principles of the abacus, small electronic calculators became pocket-size when one Integrated Circuit was able to perform all functions, batteries became rechargeable or replaceable, and liquid crystal displays began to use less energy to show the numbers in a small window.

New Words and Expressions

mainframe（尤指除外部辅助装置的）计算机
Altair [天] 河鼓二（俗称"牛郎星"）
miniaturization 小型化；微型化
prefigure 预示；预想
trillion 万亿
laptop 便携式计算机；膝上（或手提）电脑
prophetic 预言的，预言性的
nominate 提名，推荐，指导，任命
flyer 飞行器，飞行物
ferry 飞渡，空运，运送

Unit Nine   Modern Science and Technology (II)

cruise 巡航，航行
envisage 想象，设想
cellular 多孔的，蜂窝状的
cordless 无绳的，不用电线的，电池式的
cosmology 宇宙论
kidney 肾脏
escalate 升级，逐渐发展
double helix 双螺旋
breed 孕育，繁殖，育种，配种
bakelite 酚醛塑料
Teflon 聚四氟乙烯，特氟纶
Plexiglass 塑胶玻璃
terylene 涤纶
polyethylene 聚乙烯
polypropylene 聚丙烯
polymer 聚合物
nylon 尼龙
viscous 黏的
CD-ROM (Compact Disc Read-Only Memory) 只读光盘
kinescope 映像真空管
staple 订书钉；用订书钉钉住
microchip 芯片
ozone 臭氧

Notes

1. Mark I 马克一号被认为是第一台万用型计算机。它的生产者和设计者给它起名为 Automatic Sequence Controlled Calculator（全自动化循序控制计算机，缩写为 ASCC），马克一号是它的用户哈佛大学起的名字。马克一号由 IBM 的 Howard H. Aiken 设计，其特点为全自动运算。一旦开始运算便无须人为介入。人们认为"这是现代电脑时代的开端"以及"真正的电脑时代的曙光"。

2. ENIAC（Electronic Numerical Integrator and Computer）电子数字积分器与计算机是第一台通用的电子计算机，于 1946 年 2 月 14 日问世。

3. IBM System 360 mainframe 美国 IBM 公司于 1964 年推出的大型机。

4. PDP（Programmed Data Processor）程序数据处理机。

5. ARPANET（Advanced Research Project Agency Computer Network）高级研究计划署计算机网。高级研究计划署（ARPA）是美国国防部所属的一个机构，该署从 1960 年开始开发计算机通信网络，1971 年起利用其通信电路与签约大学及企业研究所的计算机相联，共享信息。后来这一网络向公众

开放，即今天的因特网（Internet）。

6. IBM Personal Computer（IBM PC）IBM 个人电脑是 IBM PC 兼容机硬件平台的原型和前身，其模型号码为 5150。

7. Orville and Wilbur Wright 奥维尔·莱特（1871—1948）和威尔伯·莱特（1867—1912），美国航空先驱。1903 年 12 月 17 日他们驾驶自行研制的固定翼飞机飞行者一号，实现了人类史上首次重于空气的航空器持续而且受控的动力飞行，被广泛誉为现代飞机的发明者。

8. Yuri Gagarin（1934—1968）尤里·阿列克谢耶维奇·加加林，苏联宇航员和苏联红军上校飞行员，他是第一个进入太空的人。

9. Concorde 协和飞机是英法两国共同研制的世界第一架超音速客机。

10. Max Planck（1858—1947）马克斯·普朗克，德国物理学家、量子力学的创始人，20 世纪最重要的物理学家之一。

11. Albert Einstein（1879—1955）阿尔伯特·爱因斯坦，美籍德国犹太裔、理论物理学家、相对论的创立者，现代物理学奠基人。

12. Stephen Hawking（1942—2018）斯蒂芬·霍金，英国剑桥大学应用数学及理论物理学系教授，当代最重要的广义相对论和宇宙论家，被称为"宇宙之王"。他证明了黑洞的面积定理，即随着时间的增加黑洞的面积不减。

13. black hole 黑洞是由一个只允许外部物质和辐射进入而不允许物质和辐射从中逃离的视界（event horizon）所规定的时空区域。

14. *A Brief History of Time*《时间简史》是由英国伟大的物理学家、黑洞理论的创立人史蒂芬·霍金撰写的一本有关宇宙学的经典著作，是一部将高深的理论物理通俗化的科普范本。

15. Enrico Fermi（1901—1954）恩里科·费米，美籍意大利裔物理学家，1938 年诺贝尔物理学奖获得者。他对理论物理学和实验物理学方面均有重大贡献，首创了弱相互作用（β 衰变）的费米理论，负责设计建造了世界首座自持续链式裂变核反应堆，发展了量子理论。

16. Alexander Fleming（1881—1955）亚历山大·弗莱明，英国细菌学家和青霉素发现者。

17. Leo Baekland（1863—1944）列奥·亨德里克·贝克兰，美籍比利时

Unit Nine Modern Science and Technology (II)

人、化学家、发明家，酚醛树脂（即塑料）的发明者。

18. Du Pont 杜邦公司是世界排名第二大的美国化工公司。在 20 世纪引导聚合物革命，并开发出了不少极为成功的材料。

19. Ziegler（1898—1973）卡尔·齐格勒，德国化学家。在聚合反应催化剂研究方面有很大贡献，并因此与意大利化学家居里奥·纳塔共同获得 1963 年诺贝尔化学奖。

20. Giulio Natta（1903—1979）居里奥·纳塔，意大利化学家，在聚合反应的催化剂研究上做出很大贡献，因此与德国化学家卡尔·齐格勒共同获得 1963 年诺贝尔化学奖。

21. Clarence Birdseye（1886—1956）克拉伦斯·伯宰，美国发明家，在液体冷冻技术领域的研究引导了冷冻快餐的推广使用。

22. King Kullen self-service supermarket "卡伦王（金卡伦）"超市是美国人迈克尔·J.卡伦（Michael J. Cullen）于 1930 年创建的第一个具有现代意义的仓储式超级市场。

23. Wallace Hume Carothers（1896—1937）华莱士·休姆·卡罗瑟斯，美国化学家，尼龙和氯丁橡胶的发明者。

24. Georg Biro 乔治·比罗，于 1943 年申请了圆珠笔专利。

25. Marcel Bich 马塞尔·比克，1950 年在法国创立比克（BIC）公司。1953 年 11 月出水流畅、书写清晰的比克圆珠笔问世，比克公司也正式宣告成立。比克圆珠笔当时是一种用完即弃的短暂商品。

26. Pincus 格雷戈里·平克斯，美国生物学家，避孕药发明者之一。1955 年在东京举行的国际计划生育联合会代表大会上，平克斯宣布了避孕药的发明。

27. Vladimir Zworykin（1889—1982）维拉蒂米尔·斯福罗金，俄国物理学家、发明家。

28. John Logie Baird（1888—1946）约翰·罗杰·贝尔德，工程师及发明家和电动机械电视系统的发明人。其他发明贡献包括发展光纤、无线电测向仪、红外线夜视镜及雷达。

29. Sir Robert Watson-Watt（1892—1973）劳勃·沃森-瓦特爵士，英

国苏格兰物理学家，雷达的发明者。他是蒸汽机发明者詹姆斯·瓦特的后代。

30. Chester Carlson 切斯特·卡尔森，静电复印技术（electrophotography），或施乐技术（xerography）的发明人。

31. John Bardeen（1908—1991）约翰·巴丁，美国物理学家，因晶体管效应和超导的 BCS 理论于 1956 年和 1972 年两次获得诺贝尔物理学奖。

32. Walter H. Brattain 沃尔特·布拉顿和约翰·巴丁、威廉·肖克利创造出了世界上第一只半导体放大器件，他们将这种器件重新命名为"晶体管"。

33. William Shockley（1910—1989）威廉·肖克利，美国物理学家和发明家，晶体管的发明者之一。他一生共获得 90 多项专利。

34. Kilby（J.S.Kilby, 1923—2005）基尔比。1958 年他接受了一项设计计算机微型组件的工作，并把电阻、电容和晶体管都制作在一块硅片上，从而发明了第一个平面型的集成电路。2000 年获得诺贝尔物理学奖。

35. Noyce（Robert Norton Noyce，1927—1990）罗伯特·诺顿·诺伊斯是仙童半导体公司（1957 年创立）和英特尔（1968 年创立）的共同创始人之一，被誉为"硅谷市长"或"硅谷之父"（the Mayor of Silicon Valley）。诺伊斯也是电子器件集成电路的发明者之一。

36. Hoff（Marcian Edward Ted Hoff, Jr., 1937— ）泰德·霍夫，微处理器（microprocessor）发明者之一。

37. Charles H. Townes（1915— ）查尔斯·汤斯，美国物理学家、教育家，1964 年获诺贝尔物理学奖。

38. Arthun Schawlow（1921—1999）阿瑟·肖洛，美国物理学家，1981 年获诺贝尔物理学奖。

39. Dr Theodore H. Maiman（1927—2007）西奥多·哈罗德·梅曼，美国物理学家，曾制造了世界上第一台激光器。著有一本名为《激光奥德赛》（*The Laser Odyssey*）的书来描述激光器的诞生。

40. Hughes Research Labs HRL 实验室，原休斯研究实验室，它是休斯飞机公司的研究机构。

Unit Nine   Modern Science and Technology (II)

Exercises

I. Find out the English equivalents to the following Chinese terms from the passage.

1. 硅片
2. 腕表
3. 年度风云人物
4. 大型喷气机
5. 超音速飞机
6. 航天飞机
7. 青霉素
8. 电动真空吸尘器
9. 聚苯乙烯树脂
10. 圆珠笔
11. 避孕丸
12. 阴极射线管
13. 空中交通管制
14. 静电复印处理技术
15. 条形码识别
16. 光纤电缆

II. Translate the following English expressions into Chinese.

1. telegraph pole
2. dial telephone
3. mobile in-car phone
4. portable cellular phone
5. radiotelephone system
6. cordless phone
7. pager
8. teleconference
9. portable fax machine
10. portable phone

III. Decide whether the following statements are True (T) or False (F).

1. The first computer in the world was the Harvard-IBM Mark I computer in 1944.
2. *Time* magazine nominated the computer as Machine of the Year in 1981.
3. In 1900, people never imagined that diagnosis could be made by laptop computer.
4. The first flight of the Boeing747 Jumbo jet can carry 400 people.
5. The Concorde, the first supersonic flight, cruises at twice the speed of sound.

6. The first mobile phone was invented by Bell Telephone in 1959.
7. Stephen Hawking is noted for his quantum theory and Einstein is noted for his theory of relativity.
8. Birdseye peas are one of the first frozen foods to be sold in small packets in the United States.
9. The ballpoint pen with metal refills was invented by Hungarian Georg Biro in 1953.
10. Radar is extensively used today in air traffic control, tracking weather patterns and spacecraft.

Unit Ten
Modern Development in China

Part A Lecture

推　理

在阅读一篇文章时，我们常常利用文字的表面含义来了解作者想表达的内容。然而，文章作者并不总是用一目了然的词句来表达思想和描写事物。有时他们出于某种原因，不是直截明了地陈述自己的观点，而是采用隐晦的、含蓄的表达方式。这就需要我们掌握逻辑推理的方法，根据事物发展的自然规律以及语言本身的内在联系，从文章的字里行间去领会、揣摩作者的言外之意，从而获取尽可能多的信息。具有获得句子或文章所表达的隐含信息的推理能力是阅读理解的高级技能之一，也是通过各类考试的必备技能。本单元讲座将详细介绍推理的技巧，并针对逻辑推理、常识推理和综合推理进行专项训练。

1 阅读推理的技巧

推理在逻辑上是思维的基本形式之一，是由一个或几个已知的判断（前提）推出新判断（结论）的过程。在一定程度上，推理与数学上的推导相似。我们把有关的文字作为已知数，即推理的前提，从中推断未知数，即推理的结论。这里，据以推理的前提是短文的有关文字部分，可能是短语或句子，也可能是若干句，甚至整段或全文。阅读下面例文并分析推理过程。

Most scientists believe that the continents of North America and Europe were joined around 200 million years ago. Since that time, these continents have been drifting apart at a rate of two inches per year. As they drifted apart, the Atlantic Ocean formed between them.

Choose the statement that can be inferred from the passage.

Unit Ten   Modern Development in China

A) The Atlantic Ocean is getting smaller.
B) The Pacific Ocean is getting larger.
C) The Atlantic Ocean is getting larger and the Pacific smaller.
D) The Atlantic Ocean is getting deeper and colder.

分析：以上四选项中，C 项是合乎逻辑的推理。虽然文中没有句子直接说明 C 项，但根据第 2 句所用的时态可以推知。大西洋在欧洲和北美洲之间的形成，是两亿年来两个大陆板块每年以两英寸的速度彼此漂离的产物，而这种漂离过程现在仍在继续。因此可以推断大西洋在日益变阔。既然大西洋变阔，太平洋必然日渐缩小。

从上例可知，正确的推理基于所读材料所提供的信息。为了做出正确的推理，我们应该注意以下几点：

1·1 了解各种推理题型的命题特点

作为一种阅读理解的题型，推理的范畴很广，涉及文章、段落、标题、某一细节、作者的创作意图、写作态度、叙述语气以及节选文章的前接后续段落（即作者的写作思路）等；这些都可作为问题提出，以测试学生的阅读推理能力。阅读推理题主要包括推测词义、推断主旨大意、推断作者倾向或态度及短文笔调、推断作者写作思路、逻辑推理和常识推理等。

1.1.1 推测词义

词义推测题要求学生对短文中个别关键词、难词进行解释。这种题型要求同学们善于捕捉所遇词上下文中的各种线索，掌握这些线索才能及时推断生词的词义。通过上下文推测词义是重要的阅读技巧，也是重要的测试技巧。推断词义的方法有利用常用的辅助词（or、that is 等）、标点符号（逗号、破折号、括号等）、生词后所带的解释或例子、生词前后的内容、生词前后语境中出现的反义词等线索以及常识等。如：

A volcano is a kind of chimney, or "vents" which goes down to a liquid deep inside the earth, called "magma". Three types of material come out of the vent: a hot liquid called lava, pieces of rock, and great quantities of gas. The lava and rock often collect round the vent and form what is known as the volcano's "cone". Volcanic eruptions vary between two extremes. In one, the lava comes quietly to the surface and flows away as a river, causing little damage except to objects directly in its path. On the other extreme great explosions occur, frequently blowing away the cone and causing great damage. The great majority of the world's volcanoes are intermediate between these two extremes.

The word "vent" means _____.

　　A) something like a chimney in the volcano

　　B) magma deep inside the earth

　　C) the volcano's cone

　　D) lava, rock, and gas

分析：根据 chimney 后的逗号及辅助词 or，我们可以猜出 vent 和 chimney 是同义词，另根据上下文 Three types of material come out of the vent，我们可以确定 vent 是"火山口"的意思。选项 A 为正确答案。

▶ 1.1.2 推断主旨大意

　　短文的主旨大意类推断题是阅读测试中考查学生阅读能力的主要方式。该题型指针对短文的主题（subject）、中心思想（main idea）、标题（title）或目的（purpose）拟题。尽管题型或用词不尽相同，但针对的都是短文的主旨大意。推断短文主旨大意的关键是确定主题句。如：

　　The Einstein observatory is an orbiting spacecraft that mapping in detail extremely distant sources of X-rays. It has sent back information that has led to a new concept of how the universe was formed. Some leading scientists now conclude that the evolution of the universe depends heavily

Unit Ten　Modern Development in China

on sequences of explosions and that the shock waves that followed them may have played an important part in galaxy.

What is the best title for this passage?
　　A) New Ideas about Einstein's Theories.
　　B) Mapping the Galaxies by Satellite.
　　C) New Ideas on the Formation of the Universe.
　　D) The Evolution of X-rays.

分析：从短文中可以看出，全文都是围绕着 information that has led to a new concept of how the universe was formed 这个中心论点来发展的，因此，第 2 句为主题句。文中第 1 句说明了这个理论的由来，而短文的后半部分阐述这个理论所带来的结果——对宇宙的形成已形成了一个新的认识。所以，短文不是关于 Einstein's theories、mapping 或 X-rays，而是关于宇宙的形成。据此推断出答案为 C 项。

▶ 1.1.3 推断作者倾向或态度及短文笔调

针对作者倾向（tendency）或态度（attitude）以及短文笔调（tone）的推理题要求学生就作者对论述对象持何种思想倾向做出判断。作者的思想倾向往往隐含在字里行间。因此，我们要依靠短文的中心思想作为推理的前提，同时注意作者的措辞。

　　Worldwide fame burst upon Albert Einstein on November 7, 1919, when British astronomers announced they found the first confirmation of Einstein's general relativity theory. Einstein had already become known in scientific circles because of his two astonishing theories: the special theory of relativity, published in 1905 when he was only twenty-six and a minor clerk in the Swiss patent office, and the general theory of relativity, advanced between 1913 and 1915. He was considered so brilliant by other scientists that in 1914 he was invited to join the prestigious Royal Prussian Academy of Sciences and to become head of the research branch at the Kaiser Wilhelm Institute. He accepted the offer and moved to Berlin.

How does the author seem to feel about Albert Einstein?

A) Uninterested   B) Critical   C) Impressed   D) Overwhelmed

分析：短文简略地记叙了爱因斯坦获得学术成就和声望的经历。尽管短文没有直接陈述作者对爱因斯坦的评论，但从某些记叙的措辞中仍可窥见作者对爱因斯坦的称颂，如：Worldwide fame burst upon（声震全球），so brilliant... that...（……才华横溢……以致……）。据此我们可以推断 C 项为正确答案。

▶ 1.1.4 推断作者写作思路

推断作者写作思路指推断短文的前接或后续段落讨论的主题。这类题要求学生推断与测试段落或短文有关的上文或下文（未作为测试部分印出）的主要论述对象或主题。因此，我们要了解一些写作的基本知识，熟悉段落之间的衔接与连贯手段，分析段落的开头与结尾部分，从中找到某种提示作为推理的前提和依据。

One hundred and thirteen million Americans have at least one bank-issued credit card. They give their owners automatic credit in stores, restaurants, and hotels, at home, across the country, and even abroad, and they make many banking services available as well. More and more of these credit cards can be read automatically, making it possible to withdraw or deposit money in scattered locations and any time, whether or not the local branch is open. For many of us the "cashless society" is not on the horizon—it's already here.

According to the passage, it is possible to tell us _____ in the next chapter.

A) how to use credit in hotels

B) when we withdraw money

C) about its defects

D) the meaning of "cashless society"

分析：本题要求对将要叙述的下一段内容做推断和预测。从本段可知：在

Unit Ten Modern Development in China

美国有上亿的人至少持有一张银行发行的信用卡。其持有者可以在商店、饭店、旅馆、家里、全国甚至国外自动信用赊款购物，并且还可享受很多银行服务。越来越多的信用卡可以自动读取，使用户可以在不同地点、时间存取钱，而不管当地银行是否开门。对我们许多人来说，"无钞票社会"并不是刚刚兴起，它早已存在。纵观全文，我们知道，本段说的都是信用卡的优点，那么下一段最有可能谈的是它的缺点。C 项是最好的选择。其实，这是一篇"优点－缺点－评论"三段论文章，了解一些写作的基本知识对推理是有帮助的。

▶ 1.1.5 逻辑推理

逻辑推理题要求学生就短文陈述的论点或描述的事实进行推理，从而得出合乎逻辑的结论。逻辑推理的关键在于对短文有关部分的透彻理解。如：

The cost of health care in the United States has increased tenfold during the last twenty-five years, tripled during the last ten years, and hospital was charging $2000. Intensive care cost more than $300. Today hospital charges for surgery and post-operative care can easily reach $10,000, including $3000 for the surgeon and $700 for the anesthesiologist. One reason that the cost is so high can be traced to advances in medical technology, which includes expensive equipment and highly-trained, highly paid personnel to run it. As an example, the CAT, a computerized scanner used for diagnostic purposes, now costs $600,000.

According to the passage, the new medical technology _____.

A) has no effect on the high cost of health care

B) contributes to the high cost of health care

C) may solve some of the problems related to the high cost of health care

D) requires less highly-trained, less highly-paid personnel

分析：短文主要论述美国近 25 年来，特别是近 10 年来，医疗费用迅速提高的情况，并分析了其产生的原因。题干要求就新的医疗技术对医疗费用迅速增长所产生的关系做出正确的判断。从短文第 4 句"医疗费用昂贵的原因之一在于现代医疗技术需要昂贵的医疗设备以及受过高级训练的、拿高薪的人员。"可以判断，只有 B 项指现代医疗技术对医疗费用高起了影响作用，符合作者的意图。

▶ 1.1.6 常识推理

常识推理题是指原文提供的信息本身无法满足解题条件，这时必须结合常识才能对答案做出抉择的题型。因此，我们在推理时，不仅要理解有关细节，还要了解有关的背景知识。

A recent investigation by scientists at the U.S. Geological Survey shows that strange animal behavior might help predict future earthquakes. Investigators found such concurrences in a ten-kilometer radius of the epicenter of a fairly recent quake. Some birds screeched and flew about wildly; dogs yelped and ran around uncontrollable.

Why can animals perceive these changes when humans cannot?
A) Animals are smarter than humans.
B) Animals have certain instincts that humans don't possess.
C) By running around the house, they can feel the vibrations.
D) Humans don't know where to look.

分析：根据短文我们知道，动物的异常行为可能有助于预测地震。为什么动物能觉察这些变化而人类则不能呢？原文并没有直接说明，但根据常识我们可判断出"动物具有人类所不具备的某些直觉"，故 B 项正确。

以上我们简单概述了推理的题型特点，下面将会针对逻辑推理、常识推理和综合推理进行专项训练。

1·2 推敲、揣测文章或句子的言下之意

从上文的几种主要推理题型中我们可以看到，每种推理题都要求我们尊重事实，严格地从短文陈述的论点和描述的事实出发进行推理，以保证得出

Unit Ten Modern Development in China

的判断有理有据、合乎逻辑。因此，我们在对文章进行推理时，一定要认真阅读材料，分析已有信息，仔细琢磨字里行间的含义，并根据短文中的相关文字部分进行推测。

▶ 1.2.1 善于分析文章中的已有信息

在阅读过程中，我们要注意收集各种线索，捕捉相关信息，做到善于分析文章中已提供信息的深层含义。且看下例。

It is all very well to blame traffic jams, the cost of petrol and the quick pace of modern life, but manners on the roads are becoming horrible. Everybody knows that the nicest men become monsters behind the wheel. It is very well, again, to have a tiger in the tank, but to have one in driver's seat is another matter altogether.

According to this passage, it suggests that troubles on the road are primarily caused by _____.

A) traffic conditions

B) the behavior of the drivers

C) the rhythm of modern life

D) animals on the road

分析：本题是根据已有的几条信息，对段落中的某一细节所做的推理。在段落中，作者谈到"指责交通堵塞、汽油的昂贵以及现代生活的快节奏固然没错，但公路上的行为正变得非常可怕。人人知道，最优雅的人一坐到方向盘后也会变成魔鬼。笼子里装只老虎当然可以，但驾驶员的座上有只老虎则完全是另一回事了"。

从这里可以看出 A 项和 C 项是原文里陈述到的细节，而文中的老虎只是一个比喻——那些莽撞的司机们，D 项也就不可能了。而原文里第二行里 but 后的那一句话用一个转折词否定前文的事实。因此，从字里行间可推断出 B 项是唯一的正确选项。

1.2.2 注意词在语境中的含义

推理要有依据,而依据往往来自文中词汇。当作者不愿或不便表明自己的态度时,便在用词上下功夫,有目的地使用某些词语,暗示自己的观点。这就加深了推理判断的难度并使之显得有些微妙。但只要我们真正理解和把握作者所用词语的内涵,也就不难做出正确的推断。如:

Herbert Hoover remained blindly passive to economic depression of his country.

分析:作者极有用意地在句中使用了 blindly 一词暗示读者 Herbert Hoover 看不见他的国家所面临的经济困难。blindly 一词的原意是指"眼睛失明,瞎的"。如果一个人眼睛失明,说明他有身体缺陷、不正常。但如果用于谴责某人眼瞎、看不见,通常指这人顽固、执拗、不听人言。可见,通过 blindly 一词,作者把自己对 Herbert Hoover 的评价含蓄地告诉了读者。其言外之意是"Hoover is blind, handicapped and stubborn."。所以说,要理解作者的真实意图,就必须留意作者的遣词造句。

1.2.3 透过表明的文字信号了解字里行间的含义

合乎逻辑的推理有赖于我们对短文的有关部分的把握、分析与理解,要求我们从字里行间悟出作者的言外之意。做推理题时,有时要靠字里行间的含义来完成推理过程。如:

When the phone finally rang, Joe leaped from the edge of his chair and grabbed for it.

分析:上句中的单词 finally 表明 Joe 可能等候这个电话有好一会儿了;leaped 和 grabbed 则含蓄地说明 Joe 的紧张和焦虑;Joe 在椅子上的位置 the edge of his chair 也表明 Joe 的不安和迫切期望。由此,我们可以认为:Joe 要等的这个电话一定非常重要。如果对上述例句中字里行间的含义不甚理解,其推断的结论亦可能会出差错。

Unit Ten Modern Development in China

推断题的答案虽在短文中无法直接找到，但推断的前提仍能（且必须）在短文中找到相应的文字部分。阅读下面的句子和段落，试分析哪几个选项可以从句子或段落中推理得到。

(1) Krill, which are the main diet of whales, have been cited as one of the world's biggest unexploited food resources.

A) Whales eat more krill than anything else.

B) The world has a number of unexploited food resources.

C) Whales are one of the world's biggest unexploited food resources.

D) The writer believes that krill constitute one of the world's biggest unexploited food resources.

分析：从句中可知，作为鲸的主要饮食的 krill（磷虾），已被引证为世界上最大的没有被开发的食物来源之一。A 选项"鲸吃磷虾比其他东西多"可以从句中推理得到，因为该句已告诉我们磷虾是鲸的主要饮食；B 选项"世界上还有很多没有被开发的食物来源"也可以推理得到，依据是磷虾是世界上没有被开发的食物来源之一，言下之意是还有很多没有被开发的食物来源；C 选项"鲸是世界上最大的没有被开发的食物来源之一"不能推理得出，句中说的是磷虾，而不是鲸；D 选项"作者认为磷虾构成了世界上最大的没有被开发的食物来源之一"不能推理得到，因为句中的被动语态 have been cited 表明是客观上被人引证的，而不是作者主观认为（The writer believes）的。因此，从句中只能推理出 A、B 选项。

(2) The Incas had never acquired the art of writing, but they had developed a complicated system of knotted cords called quipus. These were made of the wool of the alpaca or llama, dyed in various colors, the significance of which was known to the officials. The cords were knotted in such a way as to represent the decimal system. Thus an important message relating to the progress of crops, the amount of taxes collected, or the advance of an enemy could be speedily sent by trained runners along the post roads.

A) Because they could not write, the Incas are considered a simplistic, poorly developed society.

B) Through a system of knotted cords, the Incas sent important messages from one community to another.

C) Because runners were sent with the cords, we can safely assume that the Incas did not have domesticated animals.

D) Both the color of the cords and the way they were knotted formed part of the message of the quipus.

E) The quipus were used for important messages.

分析：本段意为：印加人从来没有获得写字这种艺术，但他们却发展了一种复杂的结绳记事的系统，称为结绳文字。它们由羊驼毛或骆驼毛制作而成，染成各种不同的颜色，打结和染色的重要意思官员们是知道的。绳子打结的方式体现了十进制的体系。因此，有些重要的信息，如庄稼的长势、税收金额，或敌人推进的速度等都由训练有素的跑步者沿着邮路快速地递给官员。

根据这段话，我们判断下面哪些选项可以推理得到。A选项"因为印加人没有学会写字，所以他们被认为是一种简单的、发展很差的社会"是错误的，通过段落可知，虽然他们没有文字，但通过结绳记事可以做很复杂的事，说明他们的文明很发达；B选项"通过结绳记事方法，印加人可以把重要信息从一个社区传到另一个社区"从段落中可以推理得到；C选项"因为结绳是由人跑步送的，所以我们可以完全地假设印加人没有家庭驯养的动物"是错误的，由人跑步送信息并不等于没有家养动物，因此这个推理不能成立；D选项"绳子的颜色和打结的方式都暗含着一些意思"可以从段落中推理得到；E选项"结绳文字通常用于传递或记录一些重要信息"可以推理得到，依据是最后一句话。

从以上例文可知，推测文章或句子的言下之意的技巧必须经过推理才能得到。要注意：1）进行推理判断要以句子或文章本身所提供的信息为依据，不要把自己的主观意见或所了解（但句子本身没有提供的）的事实作为推理的依据。推理的前提是在原文里有实在的依据、可靠的信息，决不可脱离原文，以主观想象进行臆测。2）推理题的答案应该是文章没有直接说明的意思，而是根据文章所陈述的事实、观点加以逻辑的延伸方可得到的结论。也就是说，选择答案时不能按事实题那样，因为推理题没有直接说出来，应该是"事实＋逻辑"，需要有一个推理的过程。

另外，凡含有原文中已被明述（state）、复述（restate）的事实和信息的选项，都不可选为答案，否则就不是推理题。推理题中常含有 infer、suggest、imply 等词。

Unit Ten　Modern Development in China

2　综合练习

　　从上一节我们知道，推理是阅读中必不可少的主要技能；正确的推理是读者根据文中的事实，通过分析和判断，推导出作者未明确陈述的、隐含的意思。正确的推理技巧应该是：1）忠于原文，切忌用自己的观点取代作者的原意，同时充分利用自己对与短文内容题材有关的背景知识的已有了解和常识；2）认真分析短文中的论据和细节，把握主题；采取谨慎和严谨的态度，尊重事实，以短文为推理依据；3）注意作者用词造句的特点和倾向，客观预测作者的立场、态度和写作意图；4）熟悉推断题选择项的命题特点，并善于在阅读中进行由表及里、由此及彼的推断，透过文字表面，洞见语外含义。

　　要真正提高推理技能，就要通过大量的阅读实践，并从实践中悟出其中的奥秘。以下我们将以逻辑推理、常识推理以及推理综合应用题型为例，训练正确推理的技能。

2·1　逻辑推理训练

　　逻辑推理题要求就短文陈述的论点或描述的事实进行推理。正确的推理建立在透彻理解短文有关部分的基础上，也就是针对题目的要求和提示，对原文中找到的所有相关信息进行仔细分析，摸清它们相互之间时间、方位、因果、对比等逻辑关系，并在此基础上进行综合推理。在推理过程中，有时某些词语的语义发挥着至关重要的作用。此外，能否准确把握某些代词的指代对象对辨清逻辑发展也很关键。

例12

　　"How do I hide the car?" Smith murmured: "It is much easier to hide a leaf in a forest than in a place without trees". He drove a new BMW, which was stolen from a millionaire. Suddenly he got a good idea. Then he turned to left and drove up the way to Chicago.

It can be inferred that he would hide the car in _____.

　　A) forest　　　B) a river　　　C) a parking area　　　D) his home

分析：这是一道典型的逻辑推理题，在阅读技能训练中十分有名，它叙述一个小偷思考如何藏匿好偷来的车。文中没有说出任何地点，却问我们是否可推理出小偷将车藏在哪里？我们来延循他的思路：

在什么地方藏车最好？

假如要藏一片树叶，哪里最好？

森林！

同样的道理，哪里藏车最不容易看出来？

停车场！

从以上思路可以得知，C项正确。

Together with earthquakes, volcanoes are phenomena which both delight and terrify the human mind at the same time. Some of the most beautiful mountains in the world, admired by all who see them, are volcanoes. On the other hand, volcanoes have throughout history caused great destruction. The term volcano is associated with the island of Vulcano just north of Sicily. In classical times, this was thought to be the home of the god Vulcan—the god of destruction. Volcanoes have always been objects of mystery, and this is true today even despite the advances of science.

The paragraph implies that in classical times _____.

A) there were a lot of volcanic eruption on the island of Vulcano

B) Vulcan lived on the island of Vulcano

C) the island of Vulcano lay, and still lies today, just north of Sicily

D) there were a lot of volcanic eruption on the island of Sicily

分析：原文主要讲述有关火山爆发时的情况以及"volcano"一词的来源，而且文中指出"In classical times, this was thought to be the home of the god Vulcan—the god of destruction."，由此得出结论，古代在"island of Vulcano"经常发生火山爆发，人们才把这个岛当作毁灭之神的家，故A项正确。

Unit Ten　Modern Development in China

2·2　常识推理训练

常识推理题是指当原文提供的信息本身不够充足时，必须结合常识才能进行推理的题型，这类试题大都针对特定的细节。正确解答这类试题，不仅取决于对有关细节的理解，而且还需依靠对有关背景知识的了解。需要特别指出的是，这种解题方法不是抛离原文而对问题纯知识性的解答。

The average population density of the world is 47 persons per square mile. Continental densities range from no permanent inhabitants in Antarctica to 211 per square mile in Europe. In the Western Hemisphere, population densities range from about 4 per square mile in Canada to 675 per square mile in Iceland to 831 per square mile in the Netherlands. Within countries there are wide variations of population densities. For example, in Egypt, the average is 55 persons per square mile, but 1300 persons inhabit each square mile in settled portions where the land is arable.

(1) There are no permanent inhabitants in Antarctica because _____.

　　A) it is too hot

　　B) it is too cold

　　C) there is no transportation

　　D) it has only recently been discovered

(2) This paragraph has probably been taken from _____.

　　A) an almance　　　　　　B) a textbook on economics

　　C) a world geography book　　D) the 1960 census report

分析：

　　题1要求回答在南极洲没有人类居住的原因。据常识，A项、C项、D项都不符合事实。故答案为B。

　　题2要求回答本篇短文可能选自何种读物。A项为某本年历，B项为某本经济学教材，C项为某本世界地理学书籍，D项为1960年的人口普查报告。据常识，A项、B项可以迅速排除。原文主要论述世界人口密度分布情况，按常识这属于地理学研究的范畴。故答案为C。

Those who despise the weather as a conversational opening seem to me to be ignorant of the reason why human beings wish to talk. Very few human beings join in a conversation in the hope of learning something new.

According to the author, what part does weather play in conversation?

A) It shows people's ignorance of purpose at conversation.

B) It can provide a topic to break the ice.

C) It indicates that very few people hope to learn anything new from conversation.

D) It can provide a topic of conversation that is acceptable.

分析：我们知道外国人交谈时习惯于以谈天气开始，而中国人则习惯于问"吃饭了吗？"，两者均是为了打破沉默，进行交谈，故正确答案为 B。

2·3 综合训练

正确的推断应建立在对短文（或短文有关部分）的透彻理解基础之上，这是正确答题的首要也是最基本的要求。短文的中心思想是推理的前提，所以首先确定短文的主题会大幅提高推理的正确率和可靠度。

Astronomers say Arizona's observatories are being jeopardized by the state's burgeoning population, because artificial lights glow steadily brighter and make the stars harder to see. Research has already been severely impaired at the Lowell Observatory in Flagstaff, Arizona, by the lights of the encircling city of 39,000. Each year, the observatory's telescopes find it more difficult to penetrate the artificial lights. As a result, astronomers are working with local officials to dim the glow with restrictions on searchlights and requirements for light shields. Mercury-based streetlights, the most common type in cities in the United States, are especially disturbing to stargazers. An alternative would be sodium lights, which emit no ultraviolet waves and free the ultraviolet band that is essential to spectral astronomy.

Unit Ten Modern Development in China

(1) According to the passage, what is causing a problem?

　　A) City lights.　　　　　　B) Local crime.

　　C) Improper shields.　　　 D) Air pollution.

(2) It can be inferred from the passage that the people possibly responsible for the problem in question are _____.

　　A) Astronomers　　　　　 B) Light bulb manufacturers

　　C) Researchers　　　　　　D) Air-polluters

(3) From the passage, which of the following can be inferred about mercury-based lights?

　　A) They can penetrate other lights.

　　B) They are extremely expensive.

　　C) They are bad for reading.

　　D) They emit ultraviolet waves.

(4) According to the passage, which of the following might contribute to the solution of the problem?

　　A) Introduction of ultraviolet lights as streetlights.

　　B) Moving of the observatories to other places.

　　C) Control of the population in the encircling city.

　　D) Use of space station as observatory.

(5) What is the main purpose of the passage?

　　A) To explain recent research.

　　B) To advertise for better telescopes.

　　C) To criticize current studies.

　　D) To report on an existing situation.

分析：短文第 1 句为主题：人口激增增加了人造照明光的亮度，给天文观察带来了困难。第 2 句举了亚利桑那 Lowell 天文台严重受影响的实例；第 3 句以天文望远镜穿透人造光的难度与年俱增的事实说明现状的严重性；第 4 句讲天文学家对此采取的行动（问题严重到非行动起来不可的地步）：与地方官员交涉以限制探照灯，使用路灯罩；第 5、6 句为第 4 句的延伸：把水银路灯换成钠灯及其理由———一种缓解目前问题的方案。

　　就问题而言，题 1 要求推断与短文主题有关的内容：本文讨论了一个问题，是什么引起了这个问题？脱离短文，根据常识，容易把 B 项或 D 项当作答案，但这两项与短文主题相去甚远；基于短文，应是 A 项或 C 项，但 C 项"路灯罩不合适"只是部分地增加了天文观察的困难，无法覆盖 search lights 和钠灯替代汞灯这些细节，故答案为 A。

题 2 推断哪些人可能对人造光太亮的问题负责；抓住了主题，不难排除 D 项，A 项和 C 项是问题的受害者。由最后两句推知：如果制造钠灯替代汞灯，能大大缓解目前的问题，这样责任就落到了 B 项"灯泡制造者"的肩上。

题 3 要求根据有关汞灯的细节做出推断，由最后一句（不释放紫外波的钠灯可作为汞灯的代用品）推断答案应是 D。

题 4 要求根据短文内容推断什么可能有助于解决目前的问题，C 项可以从主题句和第 3 句得到验证。

题 5 要求对写作目的做出推断，根据短文主题，答案为 D。

In an age when waste today means a lack tomorrow, making use of every available resource becomes more and more important. As coal is being used in greater amounts to produce electricity, larger amounts of ash, a by-product of coal, are produced.

When coal is burned in a boiler, two kinds of ash by-product are produced: a heavy bottom ash and a fine as powder fly ash that is filtered and captured by precipitators（聚尘器）. About 10 to 15 percent of the coal by-product is bottom ash, which is used like sand on icy streets and highways and also on highways as paving material.

It is the fly ash, however, that is receiving the greater amount of attention. Once considered a waste, fly ash is now classified as a natural resource in the United Stated by the state Maryland. In accordance with state and federal environmental restrictions, fly ash is placed in controlled landfills, where it is compacted and covered with soil. The seeds of various grasses and plants are then placed in the soil to make the land productive and to provide permanent storage.

Fly ash may be used as an addictive to concrete in the construction of dams, bricks, and roads, and can replace up to 20 percent of the cement used in concrete. As a by-product of burned coal, fly ash requires no additional expenditure of energy to be produced, whereas cement production requires great amounts of energy.

Unit Ten　Modern Development in China

　　Using fly ash in building materials is not a new idea. The Romans used a natural form of fly ash from volcanoes to build their roads and aqueducts, many of which are still standing.

(1) The best title for this passage would be _____.
　　A) New Uses for Bottom Ash　　　　B) The Use of Coal
　　C) Ash-Concrete Additive　　　　　D) New Uses for Fly Ash

(2) By "waste today means a lack tomorrow" (Para. 1, Line 1) the author implies, in the passage, that _____.
　　A) we should keep ash for future use
　　B) we should make good use of fly ash now for the sake of the future
　　C) we'll run out of coal very soon if we waste it too much now
　　D) we'll produce smaller amounts of ash in the future

(3) It can be inferred from the last paragraph that the Romans _____.
　　A) learned to use cement to build their roads and aqueducts.
　　B) made use of a natural form of fly ash to build their roads and aqueducts
　　C) realized that waste today meant a lack tomorrow
　　D) suffered from some volcanic explosions

分析：

　　本短文主要是关于煤燃烧后所剩下的 fly ash，过去是垃圾，现在却被看作是一种自然资源，它有新的用途。第一个用途是 fly ash is placed in controlled landfills to make the land productive，其次是 using fly ash in building materials，故题 1 正确答案为 D。

　　短文中提到，我们应充分利用一切可用的自然资源。原文主要内容是讲 fly ash 不再是垃圾，而是一种自然资源，并列举了它的两种新的用途。由此我们推断出题 2 正确答案为 B。

　　短文最后一段说，Romans 最早想到了用火山灰作为建筑材料来修建道路、沟渠，而且非常耐用，其中许多还保留至今。据此可以推断罗马时常有火山爆发，如果没有火山，就不可能有火山灰，更不可能用它来作为建筑材料。因此题 3 正确答案为 D。

　　从以上大量的推理实践可知，在平时阅读时，我们不仅要正确理解词语的表面含义，还应越过字面的限制，透过表层的意思，结合众多细节与事实，对文章进行深层次的推断，掌握隐含在文章中的语义，才能更透彻地领会作者的真实意图，从而获得更为完整的理解。

Quiz

Read each of the following paragraphs, and choose the best choice to the question.

1

The manner in which desert locust plagues develop is very complex, but the two most important factors in that development are meteorology and the gregariousness of the insect. Since locusts breed most successfully in wet weather, rain in the semiarid regions inhabited by the desert locust provides ideal breeding conditions for a large increase in population. This increase must be repeated several times in neighboring breeding areas before enough locusts crowd together to form a swarm. As the supply of green, palatable food plants decreases toward the end of the rainy season, the locusts become even more concentrated. They move on to other warm, damp, verdant places where they settle, feed, and reproduce. As this process is repeated a swarm eventually develops. Plagues are unpredictable and irregular because the meteorological patterns favorable to crowding are themselves irregular.

(1) The passage deals primarily with desert locust swarms and their _____.

A) plagues
B) reproduction
C) formation
D) unpredictability

(2) According to the passage, the locust population can increase to swarm strength within several _____.

A) months
B) breeding cycles
C) rainy seasons
D) years

(3) According to the passage, which of the following is NOT an important factor in the growth of desert locust swarms?

A) An abundant food supply.
B) Large population increase.
C) Rain in the semiarid regions.
D) The gregarious nature of locusts.

(4) According to the passage, a decrease in the number of palatable food plants causes _____.

A) a decrease in the number of locusts
B) death to the local locust population
C) a heavier concentration of locusts
D) increased breeding of locusts

(5) It can be inferred from the passage that in periods of little or no rain the locust population becomes _____.

A) smaller
B) denser
C) more active
D) more unpredictable

(6) The next paragraph would most probably deal with _____.

A) the gregarious nature of locusts
B) rain patterns in the desert
C) the type of food plants preferred by locusts
D) the control of locust plagues

2

All languages change with time. It is fortunate for us that languages change rather slowly compared to the human life span.

It would be inconvenient to have relearn our native language every 20 years. In the field of astronomy we find a similar situation.

Because of the movement of individual stars, the constellations (星座) are continuously changing their shapes. 50,000 years from now we would find it difficult to recognize Orion (猎户座) or the Big Dipper (北斗七星). But, from year to year, the changes are not noticeable.

Linguistic change is also slow, in human, if not astronomical terms. If we were to turn on a radio and miraculously receive a broadcast in our "native language" from the year 3000, we would probably think we had turned in some foreign-language station. Yet we hardly notice any change in our language.

Where languages have written records it is possible to see the actual changes that have taken place. We know a lot about the history of the English language, because about 1000 years of English is preserved in writing. Old English, spoken in England at the end of the first millennium (一千年), is scarcely recognizable as English. There are college courses in which Old English is studied in the same way as any foreign language such as French or Swahili (斯瓦希里语).

(1) The main idea of the passage is that language _____.

    A) bears a close relationship to astronomy

    B) evolves slowly, but continuously

    C) gradually becomes foreign to its speakers

    D) varies from year to year

(2) According to the passage, Old English is similar to Swahili in that _____.

    A) both have a similar grammatical structure

Unit Ten　Modern Development in China

　　B) both are popular college courses

　　C) both are foreign to the average American

　　D) neither is spoken any more

(3) It can be inferred from the passage that the Big Dipper _____.

　　A) hasn't changed at all in the last millennium

　　B) is difficult to recognize today

　　C) will not be visible by the turn of the century

　　D) was not recognizable 50,000 years ago

(4) Which literary device does the author employ in discussing language?

　　A) Analogy.　　　　　　B) Humor.

　　C) Sarcasm.　　　　　　D) Redundancy.

(5) Why does the author put "native language" in quotation marks?

　　A) It is an invented term.

　　B) He is quoting an authority.

　　C) Languages always outlive the people that use them.

　　D) He is using the term ironically.

Part B Reading

Modern Development in China [1]

Science and technology in the People's Republic of China has in recent decades developed rapidly. The Chinese government has placed emphasis through funding, reform, and societal status on science and technology as a fundamental part of the socio-economic development of the country as well as for national prestige. China's scientific and technical achievements have been impressive in many fields. Now China is increasingly targeting indigenous innovation and aims to reform remaining weaknesses.

In 1900, China had no modern science and technology at all—fewer than 10 people in all China understood calculus. Now in the early 21st century, the high-technology research and development gap between China and the world's advanced countries has visibly shrunk; 60 percent of technologies, including atomic energy, space, high-energy physics, biology, computer, information technology and robot, have reached or are close to the world advanced level.

The successful launches of manned spacecraft in 2003 and 2005, as well as the launch of the moon probe satellite, marked a leap in Chinese astronautics. In September 2008, China launched the manned Shenzhou VII, and a Chinese astronaut successfully made China's first-ever space walk. China became the world's third country to master the technologies required for space walks. According to China's moon probe plan, China will launch unmanned probes to the moon before 2010, and gather moon soil samples before 2020.

The amount of expenditures on research and development (R&D) activities was worth 457.0 billion yuan in 2008, up 23.2 percent over 2007, accounting for 1.52 percent of the country's GDP. Of this total, 20 billion yuan was appropriated for fundamental research programs. A total number of 922 projects under the National Key Technology Research and Development Program and 1205 projects under the Hi-tech Research and Development Program (the "863 Program") were implemented.

The year 2008 saw the establishment of seven new national engineering research centers and 51 national engineering laboratories. The number of state validated enterprise technological centers reached 575 by the end of the year. The technology centers at the provincial level numbered 4886. 828,000 patent applications were made both home and abroad, of which 717,000 were domestic ones, accounting for 86.6 percent of the total. A total number of 290,000 patent applications for new inventions were accepted, of which 195,000 were from domestic applicants or 67.1 percent of the total.

By the end of 2008, there were altogether 24,300 laboratories for product inspection, including 376 national inspection centers. There were 170 organizations for product certification and management system certification, which accumulatively certified products in 38,000 enterprises. A total of 3071 authorized measurement institutions enforced compulsory inspection on 41.90 million measurement instruments in the year. A total of 6373 national standards were developed or revised in the year, including 2714 new standards. There were 1314 seismological monitor stations and 31 seismological remote monitoring network stations. Oceanic observation stations numbered 67 and oceanic monitoring spots reached 9200. Mapping departments published 1834 maps and 309 mapping books.

As China's largest scientific and technology project of the 20th century since the reform and opening-up policies were introduced, the national Key Science and Technology Project was approved at the Fifth Meeting of the Fifth National People's Congress on November 30, 1982. The project's goal is to solve long-lasting problems of overall importance related to agriculture, electronic information, energy resources, transportation, materials, resources survey, environment protection, medical care and sanitation, etc., in national economic and social development. So far, it remains the largest science and technology undertaking that produces the most widespread influence on the national economy.

The implementation of the Projects has played a significant role in the sustainable development of China's economy, science and technology, national defense and society. During the 10th Five-Year Plan (2001–2005) period, through the Project 210 important and significant items were arranged in the eight fields of agriculture, information, automation, materials, resources and transportation, resources and environment, medical care and sanitation, and public services. And during the period of the 11th Five-Year Plan (2006–2010), stress is given to the R&D of public service technology, key general technology, and enterprise participation to establish the mechanism of the integrating manufacturing with research; strong support is provided to industries in the fields of energy resources, natural resources, environment, agriculture, medicine and sanitation.

China has developed sci-tech cooperation relations with 152 countries and regions, signing intergovernmental sci-tech cooperation agreements with 96 of them, and joined more than 1000 international sci-tech cooperation organizations. Non-governmental international sci-tech cooperation and exchanges remain active. The China Association for Science and Technology [2] and its affiliated organizations have joined 249 international scientific and technological organizations, and more than 250 CAS scientists have held posts in international scientific organizations. The China National Science Foundation [3] has concluded cooperative agreements and memoranda with their counterparts in 36 countries.

The Chinese Academy of Sciences, as well as a dozen famous universities and colleges, including Peking University and Tsinghua University, are located in the Zhongguancun area in Haidian District, Beijing. Boasting a dynamic economy based on knowledge and information industries, Zhongguancun employs hundreds of thousands of professionals averaging

Unit Ten　Modern Development in China

about 30 in age. It is thus popularly known as the "Silicon Valley of China".

Zhongguancun is a product of the implementations of the reform and opening-up policy and the development of market economy in China. Starting in the late 1980s, the Chinese government decided to focus its attention on economic development. On October 23, 1980, Chen Chunxian, a researcher at the Chinese Academy of Sciences, established a technological development service shop under the Beijing Society of Plasma Physics in Zhongguancun. It was the first civilian-run scientific and technological business in the area.

The Zhongguancun Scientific and Technological Garden was established in June 1999. It was the first state-level hi-tech industry development zone ever founded in China.

New Words and Expressions

indigenous 本土的，土生土长的
calculus 微积分
expenditure 支出，消费，花费，使用
validate 使生效；使合法化；批准；证实
certified 被证明了的，有保证的
affiliate 接纳……为会员（或分支机构）
memoranda 买卖通知书
plasma [物] 等离子体，等离子区

Notes

1. 本文选自中国日报网，2016年6月15日。

2. China Association for Science and Technology 中国科学技术协会（CAST）是中国科学技术工作者的群众组织，由全国学会、协会、研究会和地方科协组成，组织系统横向跨越绝大部分自然科学学科和大部分产业部门，是一个具有较大覆盖面的网络型组织体系。1958年9月，经党中央批准，全国科联和全国科普合并成立中国科学技术协会。中国科学技术协会的全国领导机构是全国代表大会和它选举产生的全国委员会，全国代表大会每五年举

行一次。

3. China National Science Foundation 中国国家自然科学基金是为鼓励自然科学创新与发展而设立的基金项目。

Exercises

I. Find out the English equivalents to the following Chinese terms from the passage.

1. 载人飞船
2. 月球探测卫星
3. 太空漫步
4. 国内生产总值
5. 国家科技攻关项目
6. 国家高技术研究发展计划（863 计划）
7. 产品检验
8. 地震监测站
9. 海洋观测站
10. 改革开放政策
11. 全国人民代表大会
12. 医疗保健和卫生设施
13. 可持续发展
14. 五年规划
15. "十一五"发展规划期间
16. 科技合作
17. 中国科学院
18. 中国硅谷
19. 市场经济
20. 等离子体物理学

II. Answer the following questions by using the sentences in the passage or in your own words.

1. What marked a leap in Chinese astronautics?
2. When did China's first-ever space walk take place?
3. How many projects were implemented under the Hi-tech Research and Development Program?
4. How many patent applications for new inventions were accepted from domestic applicants in 2008?
5. How many seismological remote monitoring network stations were there by the end of 2008?

Unit Ten   Modern Development in China

6. What's the aim of the national Key Science and Technology Project?
7. During the period of the 11th Five-Year Plan, what fields is emphasis given to?
8. How many international sci-tech cooperation organizations has China joined?
9. Where is the "Silicon Valley of China"?
10. What is Zhongguancun well-known for according to the passage?

Part C Extended Reading

Yuan Longping—Father of Hybrid Rice

The Food and Agriculture Organization of the United Nations declared 2004 the International Year of Rice—the main staple food in over 30 countries in Africa, Asia, North America, and South America, and the Pacific region. Rice provides one-fifth of the world's dietary energy; by contrast, wheat supplies 19 percent and maize, 5 percent.

To honor the FAO's celebration of this crop, crucial to feeding and nourishing the world, the 2004 World Food Prize was given to two rice scientists who, working independently, each made miraculous breakthroughs that bettered the lives of countless human beings throughout the world. The 2004 World Food Prize Laureates were Professor Yuan Longping, director-general of the China National Hybrid Rice Research and Development Center in Hunan, China, and Dr. Monty Jones of Sierra Leone, a former senior rice breeder at the West Africa Rice Development Center and presently executive secretary of the Forum for Agricultural Research in Africa in Accra, Ghana.

Professor Yuan Longping and Dr. Monty Jones of Sierra Leone

Unit Ten   Modern Development in China

For his breakthrough achievement in developing the genetic materials and technologies essential for breeding high yielding hybrid rice varieties, Professor Yuan Longping was awarded the World Food Prize in 2004. He is considered the first scientist to successfully alter the self-pollinating characteristics of rice and facilitate the large-scale production of hybrid rice, which has 20% more yield than elite inbred varieties. He took the unknown path in achieving the rice hybrid miracle.

Professor Yuan, who became known as "Father of Hybrid Rice", was born in Beijing in 1930. His education at the Southwestern Agricultural College in Chongqing, majoring in agronomy, marked the beginning of his lifelong work in agriculture. Upon his graduation in 1953, he took a teaching job at the Anjiang Agricultural School in Hunan Province. Along with teaching, he also conducted scientific experiments involving asexual crosses between crops. These experiments led him to the conclusion that there were faults in the accepted breeding approach, and he then began concentrating on experiments based on the genetic theories of Mendel & Morgan, which were different from traditional theories.

Professor Yuan began his research on developing hybrid rice in 1964 at a time when it was widely accepted that hybrid vigor—or heterosis—could not be bred in a self-pollinated crop like rice, and no solutions for high-yielding hybrid seed production in self-pollinated crops were on the horizon. Nevertheless, Professor Yuan believed that heterosis is a universal phenomenon and rice is no exception. After nine years of research, he succeeded in breeding unique genetic tools, which consisted of a three-line system: male sterile line; maintaining line; and restore line—or A, B, R line, essential for developing hybrid rice. His varieties were put into commercial production in China in 1976.

He has continued his scientific exploration to develop new approaches to enhance the heterosis level and to simplify the methodology for hybrid rice breeding. In the 21st century, as the world's rice production is called upon to meet the demand of increasing population and potential decreases in planting area, Professor Yuan has conducted a careful analysis of the yield-limiting factors and breeding technologies available. In that regard, he has led a project to develop "super hybrid rice", which has an additional yield increase potential of 20%.

The additional improvements in hybrid rice breeding and production techniques have contributed greatly to increasing China's total rice output. In 2012, the area under hybrid rice has expanded to 16 million hectares, reaching about 57% of paddy, which contributes 65% of total rice output. The average yield of hybrid rice is 7.2 t/ha, while other inbred varieties yield 5.9 t/ha. It is estimated that approximately 70 million more people annually in China can be fed by planting hybrid rice, thus it helps China solve food shortage challenges successfully and provides additional income to thousands of farmers.

Professor Yuan's new hybrid rice technology has not only benefited China, but also has been enthusiastically adopted in other countries. Professor Yuan introduced Chinese hybrid rice to the world in 1979 at an international conference sponsored by the International Rice Research Institute in the Philippines (IRRI). The following year, IRRI restored its own hybrid rice research. In light of Chinese success, many countries, institutions and commercial companies started their own hybrid rice research. The United Nations' FAO made hybrid rice the first choice of its program to increase grain production outside China, and appointed Professor Yuan as the chief consultant.

He and his research associates have traveled to India, Vietnam, Myanmar, Bangladesh, Sri Lanka and the United States to provide advice and consultation to rice research personnel. Professor Yuan's research institute has trained over 3000 scientists from more than 50 countries. Farmers around the world have benefited from his techniques as hybrid rice spread throughout Asia, Africa, and the America.

Professor Yuan has published more than 60 articles and 6 monographs including Hybrid Rice Breeding and Cultivation and Technology of Hybrid Rice Production (published by FAO). His work has greatly influenced other research fields, such as plant sciences, agriculture and applied biotechnology. In recognition of his work, he has been bestowed numerous awards and honors, which include the 1981 first Special-class National Invention Prize, the 2000 National Supreme Scientific and Technology Award, the 1987 UNESCO Science Prize, the 2004 World Food Prize and the 2004 Wolf Prize in agriculture, and also in 2007 he was named a foreign associate of the National Academy of Sciences in the United States.

New Words and Expressions

dietary 饮食的；规定食物的
maize 玉米
miraculous 奇迹般的，不可思议的
laureate 戴桂冠的人；（由于艺术或科学上的成就）获得荣誉者
breeder 育种人
variety [生]变种；品种
self-pollinate （使）自花授粉
inbred 先天的；选种产生的；[生]近交的

agronomy 农艺学
asexual 无性生殖的
vigor 活力，精力，力量
heterosis 杂种优势，异配优势
sterile 不生育的；不结果实的
paddy 米；稻田
monograph 专题文章，专题著作
bestow 把……赠与，把……给予

Exercises

I. Find out the English equivalents to the following Chinese terms from the passage.

1. 联合国粮食及农业组织
2. 主食
3. 2004 年度世界粮食奖获得者
4. 杂交水稻之父
5. 孟德尔和摩根的遗传学理论
6. 科学探索
7. 育种技术
8. 首席顾问
9. 生物工艺学
10. 联合国教育、科学及文化组织

II. Translate the following sentences into Chinese.

1. For his breakthrough achievement in developing the genetic materials and technologies essential for breeding high yielding hybrid rice varieties, Professor Yuan Longping was awarded the World Food Prize in 2004.

2. His education at the Southwestern Agricultural College in Chongqing, majoring in agronomy, marked the beginning of his lifelong work in agriculture.

3. Nevertheless, Professor Yuan believed that heterosis is a universal phenomenon and rice is no exception.

4. It is estimated that approximately 70 million more people annually in China can be fed by planting hybrid rice, thus it helps China solve food shortage challenges successfully and provides additional income to thousands of farmers.

5. Farmers around the world have benefited from his techniques as hybrid rice spread throughout Asia, Africa, and the America.

Unit Eleven

Scientific Discoveries and Inventions

Part A Lecture

结　论

　　人们在阅读一篇文章后，为了获得较高的阅读效率，常常会对文中某些事实或论据进行归纳总结。有时，文章作者会直接给出结论，让读者一目了然。然而，在更多的情况下，作者在文章里没有直接给出结论，这就需要读者按照逻辑发展的规律，对文章阐述的事实或细节进行分析和概括，对某些内容进行推理，引出逻辑结论。得出正确的结论是一项重要的阅读技巧。只有在事实充分、理解透彻和仔细推敲的基础上，才有希望得出正确的结论。

1 得出正确结论的技巧

　　在阅读理解中得出正确结论，是指读者在阅读过程中，对文章中所提到的观点、态度、情绪和细节进行进一步的深加工，并在此基础上正确把握作者的观点和倾向。得出结论包括三个方面的含义：1）抓住作者的结论；2）读完一篇短文后得出正确而必然的结论；3）按指令就文章的有关内容做出合乎逻辑的结论。如：

　　The male and female mosquitoes make an odd couple. The female is a vampire and lives on blood. The male is a vegetarian that sips nectar and plant juices. Females of different species choose different hosts on which to dine. Some feed exclusively on cattle, horses, birds, and other warm-blooded creatures. Some favor cold-blooded animals. Still others prefer man.

　　While the female's menu varies, her bite remains the same. She drives her sharp, tubular snout（尖嘴）through the skin, injects a fluid to keep the blood from clotting, and drinks her fill, which takes a minute or less. It is the fluid she injects that carries disease. After her blood meal, she rests while

Unit Eleven Scientific Discoveries and Inventions

her eggs develop. She then looks for a moist or flooded place to lay them.

(1) It can be concluded from the paragraph that the male mosquito is _____.

    A) dangerous   B) relatively harmless

    C) irritating   D) aggressive

(2) Female mosquitoes _____.

    A) are a threat to most forms of living creatures

    B) cannot be controlled effectively

    C) render important service to mankind

    D) are a necessary element in the balance of nature

(3) Which of the following are natural breeding places for mosquitoes?

    A) High, dry terrain.   B) Rivers and oceans.

    C) Damp, swampy areas.   D) Mountain forests.

分析：

    根据文中第 3 句（The male is a vegetarian that sips nectar and plant juices.），我们可以得出结论：雄蚊子相对来说对人类无害。因此，题 1 正确答案为 B。

    根据文中第 2、4 和 5 句（The female is a vampire and lives on blood. Some feed on... warm-blooded creatures. Some favor cold-blooded animals. Still others prefer man.），我们可以得出结论：雌蚊子对多数动物构成威胁。题 2 正确答案为 A。

    根据文中最后两句，可以推知蚊子的主要滋生地为潮湿的地方，故题 3 正确答案为 C。

从上例可知，结论是建立在文章客观事实的基础上的。为了能正确得出结论，以下我们先了解结论与推理的异同，然后再分析得出结论的常用技巧。

1·1 结论与推理的异同点

    一方面，结论不同于推理。结论是从推理的前提中推导出来的判断，是综合了各种信息而得出的一个总论。而推理是用已有的信息内容来推断出与之有关的但没有直接说出的事情，更隐喻、含蓄的思想观点等。另一方面，结论与推理一样，都包含有一个推理的过程。推理的前提是在原文里有实在的依据和可靠的信息，绝不可能脱离原文，更不能以主观想象进行臆断。且看下面一例。针对这篇短文，提问方式的不同可以辨别推理和结论的不同。

　　One hundred and thirteen million Americans have at least one bank-issued credit card. They give their owners automatic credit in stores, restaurants, and hotels, at home, across the country, and even abroad, and they make many banking services available as well. More and more of these credit cards can be read automatically, making it possible to withdraw or deposit money in scattered locations and any time, whether or not the local branch is open. For many of us the "cashless society" is not on the horizon—it's already here.

(1) According to the passage, it is possible to tell us _____ in the next chapter.

　　A) how to use credit in hotels

　　B) when we withdraw money

　　C) about its defects

　　D) the meaning of "cashless society"

(2) According to the passage, we believe that the credit card enables its owner to _____.

　　A) withdraw as much money from the bank as he wishes

　　B) cash money wherever he wishes to

　　C) enjoy greater trust from the storekeeper

　　D) obtain more convenient services than other people do

分析：

　　题1是推理题，要求在了解全文的基础上，对将要叙述到的内容所做的一种推断和预测。纵观全文，说的都是信用卡的优点，那么下一段最有可能谈的是它的缺点，故答案为 C。

Unit Eleven　Scientific Discoveries and Inventions

　　题 2 要求在通读全文后对信用卡的种种表现做个结论，到底信用卡对其持有者有何帮助？无论 A 项、B 项还是 C 项，它们或是文章中已有的事实，或只是其中叙述到的某一点或某一方面。其实，文中谈到信用卡的作用都是为了给持有者以方便，故答案为 D。

1·2　得出正确结论的常用技巧

　　得出正确的结论既需要以短文事实为依据，也需要我们充分利用自己已有的知识经验和一般常识；既需要先抓住短文的主题，顺藤摸瓜，也需要领会主要细节与短文主题的关系和作者的写作意图。得出正确结论的常用技巧有：抓住文章的中心思想；对文章中的事实细节筛选评估；参考个人经验。

▶ 1.2.1 抓住文章的中心思想

　　一般来讲，作者在文章中所要表达的主要思想常出现在文章的主题段（往往是第一段），而其他段落基本上是围绕这一中心思想来展开的，最后再进行总结。那么，我们如果要就某一事实或现象得出正确的结论，最简单的办法就是仔细阅读文章的第一段和最后一段，抓住文章的中心思想，再进行总结概括。如果我们在阅读理解中所得出的结论与文章的主题段和总结段所表达的内容有冲突，那么，我们应对自己的结论重新审视。

　　Soccer is played by millions of people all over the world, but there have only been a few players who were truly great. How did these players get that way—was it through training and practice, or are great players "born, not made"? First, these players came from places that have had famous stars in the past—players that a young boy can look up to and try to imitate. In the history of soccer, only six countries have ever won the World Cup—three from South America and three from Western Europe. There has never been a great national team—or a really great player from North America or from Asia. Second, these players have all had years of practice in the game. Aifredo Di Stefano was the son of a soccer player, as was Pele. Most players begin playing the game at the age of three or four.

　　Finally, many great players come from the same kind of neighborhood—

a poor, crowded area where a boy's dream is not to be a doctor, lawyer, or businessman, but to become a rich, famous athlete or entertainer. For example, Liverpool, which produced the Beatles, had one of the best English soccer teams in recent years. Pele practiced in the street with a "ball" made of rags. And George Best learned the tricks that made him famous by bouncing the ball off a wall in the slums of Belfast.

  All great players have a lot in common, but that doesn't explain why they are great. Hundreds of boys played in those Brazilian streets, but only one became Pele. The greatest players are born with some unique quality that sets them apart from all the others.

According to the author, which of the following statements is TRUE?

  A) Soccer is popular all over the world, but truly great players are rare.

  B) Millions of people all over the world are playing soccer, but only six countries have ever had famous stars.

  C) Soccer is played by millions of people all over the world, but only six countries from South America and Western Europe have ever had great national teams.

  D) All over the world soccer is one of the most popular games, but it seems least popular in North America or Asia.

分析：这篇文章的第一段开门见山地提出了文章的主要思想，该句的转折句"…but there have only been a few players who were truly great"与A项中的"…but truly great players are rare"这一意义相吻合。我们如再阅读最后一段，它也在说"Hundreds of boys played in those Brazilian streets, but only one became Pele."，这与第一段的内容也符合，所以正确答案应为A。

▶ 1.2.2 对文章中的事实细节筛选评估

  有时候一篇文章是以故事发展的情节为线索展开的，所以我们很难读出它的主题段甚至结论段。在这种情况下，如果想得出正确结论，就必须对文中所列的事实细节进行筛选评估，对比总结，了解什么是重要信息、为什么重要以及每个细节之间怎么互相影响，以及一个现象的发生怎样导致另一个现象的发生等。所以简单地把文章的信息细节收集起来是远远不够的，我们

Unit Eleven Scientific Discoveries and Inventions

必须认真理解并思考它们究竟要表达什么含义。如果我们能这样阅读的话，就可以从作者告诉我们的事实中得出一系列正确的答案。

In 1931 professor and Mrs. W. N. Kellogg became the first American family to raise a chimpanzee and a child together. The Kelloggs brought into their home Gua, a seven-month old chimpanzee, who stayed with them and their infant son Donald for nine months. No special effort was made to teach Gua to talk; like the human baby she was simply exposed to a speaking household. During this period, Gua came to use some of her natural chimpanzee cries rather consistently; for instance she used her food bark not just for food but for anything else she wanted. Although Gua was rather better than Donald in most physical accomplishments, unlike Donald she did not babble and not learn to say any English words.

In the 1940s psychologists Catherine and Keith Hayes set out to improve upon the Kelloggs' experiment by raising a chimpanzee named Viki as if she were their own child. They took her home when she was six weeks old, and she remained with them for several years. The Hayeses made every effort to teach Viki to talk; they had assumed that chimpanzees were rather like retarded institutionalized children and that love and patient instruction would afford Viki the opportunity for optional language development. After six years of training, Viki was able to say four words: "mamma", "pappa", "cup", and "up". She was never able to say more, and the words she did say were very difficult to understand: in order to pronounce a p, she had to hold her lips together with her fingers.

We can conclude from the passage that _____.
  A) chimpanzees can learn to speak as humans do, when given explicit instruction
  B) chimpanzees cannot learn to speak as humans do, even when given explicit instruction
  C) chimpanzees can learn to speak as humans do when they are raised together with children when they are very young
  D) no matter what efforts are made, chimpanzees can never learn any human language

分析：这篇文章从结构上来看，没有一个很明显的具有概括性的主题段和结论段，只是讲述了两个事例。如果这时对文章中的事实作如下对比：两个实验分别是什么时候做的？谁做的实验？调查者采用什么样的手段？结果怎样？调查想证明什么？还有，这两个实验有什么相同点与差异，等等。这时，我们就不难得出结论：我们即使对猩猩进行和人类同样明确清楚的指令教育，它也不会具有像人类那样的语言功能。所以答案为 B。

▶ 1.2.3 参考个人经验

文章中的事实细节固然重要，但我们应该知道这些细节往往也来自于生活。所以我们在阅读理解中碰到的有些问题只要与个人的生活经历有关，或者能与公共的一般标准稍加联系对比，就可以得出正确的结论。这也是处理一些与我们日常生活相关的一类主题的阅读理解时，可供选择的一条捷径。

When taking part in an interview for a job there are certain approaches which can help a candidate to make a good impression.

Firstly, the candidate should normally avoid straight "yes" or "no" answers, particularly when answering negative questions. For example, if the interviewer asks, "You don't have any experience of marketing, do you?" it is better not to reply, "No, I'm afraid I don't." Much better is an answer like, "I like the challenge of a new job. I can satisfy most of your requirements and I wish to increase my experience."

The most difficult questions are those that can reveal a weakness both in the candidate's ability to do the future job and in their personality and psychology. An example of such a question would be, "I understand you lost your last job." Again, rather than a simple "Yes", the candidate could say "Over the recent past quite a few jobs have gone from my company because of the economic situation. However, I see this as a great opportunity to use my skills in such a dynamic company as yours". By answering in this way the candidate shows that it was not because of being a bad worker that they lost the job. It was also shows that he/she is viewing the situation positively.

Undoubtedly the key to performing well in an interview is good preparation, for it is obviously extremely difficult to give good answers

Unit Eleven Scientific Discoveries and Inventions

to questions that are unexpected. And, in fact, most "difficult" interview questions are predictable because they occur very frequently. A typical example of this kind of common, but tricky, question is, "Tell me about yourself." When answering this question, one shouldn't start by giving a detailed life history—the interview is not interested in your family background or where you go for your holidays. Rather interpret this question as meaning "Tell me why I should employ you". Show the interviewer why you are the right candidate by stressing your recent relevant experience. Talk on this for perhaps one minute and then explain how you feel this experience could be used beneficially in their organization.

Other difficult questions which are often asked are, "why does this job interest you?" "What is your greatest weakness?" and "Do you have an aggressive management style?" All these questions do demand very thoughtful and skillful answering and this obviously cannot be done unless there has been proper pre-interview preparation.

Thus the choice when you attend an interview is clear: you can go in blindly and hope for the best or you can prepare yourself carefully for predictable "tricky questions". Take your pick.

Which of the following conclusions can be drawn?
A) Questions from the interviewer should all be answered honestly otherwise you're unlikely to be employed.
B) Questions from the interviewer should be treated thoughtfully and answered skillfully.
C) You needn't prepare yourself carefully for an interview because the interviewer often asks some unexpected questions.
D) Taking part in an interview for a job is really something very easy.

分析：从文章的主题段（第一段）及各段的主题句中我们了解到，这是一篇关于求职面试技巧的文章。而求职面试是每个大学生都会碰到的经历，这种经历有时是直接的，有时是间接的。总之，这是我们非常熟悉又非常关注的话题，很多学校甚至有类似的选修课。另外，大众媒体对此方面的报道也很多，所以我们完全可以根据自己的经历或已掌握的信息得出如下结论：求职面试其实大有学问，既需要事先的精心准备，又要求沉着冷静、机智的临场发挥，要想在面试中大获全胜绝非易事。因此，正确答案选 B。所以我们说，生活处处有学问，关键是平时的学习积累，这一点是非常重要的。

2 综合练习

从上节的技巧分析我们知道，通过归纳推理，获得可以被已知事实证明正确的推断结果为结论。我们可以从一句话、一个段落或更长的语段得出结论。正确的结论需要充分的事实论据、周密和严谨的推理逻辑，因为只有当结论完全符合短文中相关部分所提供的事实或论据，这些事实或论据又足以证明该结论的必然性而且事实、论据本身也确凿无误时，所得出的结论才是正确可靠、站得住脚的。得出正确结论的一般步骤可以总结如下：1) 阅读全文，确定主题，抓住重要细节；2) 理解并分析这些细节和短文主题及其关系；3) 根据短文的主题和文中的事实、论据得出结论；4) 验证结论的可靠性，该结论是否基于短文提供的论据并与主题相吻。

以下我们通过短文实例分析如何得出正确的结论。

The real problem with pollution is people—the way people think about their environment. We are all reluctant to accept the fact that our natural resources are fixed—fixed, in fact, since the earth was created. We want to go on using virgin materials. We aren't educated to reusing resources, or even placing a value on "waste" products. We're a crisis society. Currently, we are fearful of losing the use of our water and air to pollution. In one context, it's probably a good thing that we are so concerned because now we'll begin to adjust our thinking on the values of natural resources and reuse. We once thought of water and air as free. They're not, not any more than the land is free. People haven't wanted to be educated on the part they must play in solving our environmental problems.

(1) The environmental crisis can be turned around and solved if people _____.

   A) insist on using virgin materials

   B) change their basic attitudes

   C) adjust to shortages and high prices

   D) clean up their rivers and lakes

Unit Eleven Scientific Discoveries and Inventions

(2) An important conclusion which can be drawn from the paragraph concerns _____.

   A) mass education

   B) government financing

   C) political responsibility

   D) strict controls

(3) It can be concluded from the statement, "We're a crisis society", that _____.

   A) Americans are pessimistic

   B) America is a warlike nation

   C) pollution is here to stay

   D) Americans must be provoked to action

(4) Which sentence summarizes the conclusion concerning the solution to our environmental problems?

   A) The first sentence.   B) The second sentence.

   C) The third sentence.   D) The last sentence.

(5) It CANNOT be inferred from the passage that _____.

   A) people are responsible for pollution

   B) the water and air around us are being polluted

   C) land is as limited as air and water

   D) the natural resources stay the same as when the earth was created

分析：本短文第 1 句为主题句：污染问题主要责任在人，在于人对待环境和资源的态度和认识。第 2 至 4 句分述了人们对环境和资源的认识问题：第 2 句，人们不愿面对这样的现实——地球的资源是一个定量，用一点就少一点；第 3 句，人们只知道一个劲儿地使用原始资源；第 4 句，人们尚未认识到资源的再利用即"废品"的价值。由此得出一个必然的结论（第 5 句）：我们的社会正处于生死存亡的关头。因此（第 6 句）人们在担心，由于污染他们会失去水和空气；而在一定意义上，这种担心是好事（第 7 句），至少人们正在改变对自然资源及其再利用价值的认识；其中的一个认识是（第 8、9 句），原以为水和空气是取之不尽的，但事实是水和空气像土地一样有限。因此最终的结论是（最后一句）：人类在保护环境方面究竟应扮演什么角色，还有待于全民教育的宣传启迪，只有每个人都知道保护环境的重要性，都自觉地为环境保护出力，环境问题才能从根本上得以解决。

题 1 为推论题，由短文主题句和结论句推知：环境危机的扭转关键在人的态度和认识的改变，答案为 B。

题 2 为结论题，由题 1 进一步推知，有关环境问题的结论必然与 A 项"全民教育"有关，B 项、C 项、D 项都不是由本文所能推导的结论，因此答案为 A。

题 3 是有关结论句"We're a crisis society"的结论：既然由于人的认识态度，我们这个社会被污染问题推到了生死存亡的关头，最佳答案应是立即行动起来，方能扭转危机、转危为安，答案为 D。

题 4 要求确定短文中哪一句概括了解决环境问题的结论，B 项、C 项显然不是答案，A 项、D 项都论及环境的关键，但 A 项（主题句）侧重起因——人的认识问题，D 项则侧重解决该问题的方案——全民教育，因此答案为 D。

题 5 是推论题，答案为 D。由主题句推知 A 项，由第 6 句推知 B 项，由第 8、9 句推知 C 项。

在对文章进行总结时，也可根据文中的相关字、词、句、段落进行推测。

Long before explorers and colonists landed on American shores, tribes of Indians named areas of land for the people who lived there. They also named the mountains, rivers or natural land marks. Indians are known for their love of nature and the poetic way they described it. Today over half of the states carry names with Indian meanings. Many were first given to rivers then later used to name the state.

Explorers, colonists, pioneers and settlers named some states in honor of kings, queens, and places in the Old World. Some names are so ancient; history cannot tell us when and how they originated.

It can be concluded from the passage that _____.

　　A) early explorers named the land

　　B) America's history is confused

　　C) Indians resented foreigners

　　D) Indians have a way with words

分析：原文的第一段叙述了印第安人和他们所命名的山脉、河流、大川。接着第二段又叙述了有一部分地名是早期的探险家和殖民者为纪念国王、女王而起的。看似前后两部分说的不是一回事，实际上都是为了阐述一个

Unit Eleven   Scientific Discoveries and Inventions

共同的主题：Indians are known for their love of nature and the poetic way they described it. 从所给的选择项看，B 项和 C 项与原文完全不符，可以排除。那么 A 项和 D 项选择哪一个呢？在这种情况下，我们还得从"主"与"次"的分析入手。如果主次分析仍不明确，可以着眼于两者在文中所占的篇幅。一般来说，作者笔墨花得多的是主要信息。由此可以推断正确答案是 D。

有些阅读理解的文章较长，加上难度稍大、时间紧等因素，读者想在阅读完后直接进行答题往往有些困难。这时就可以从问题直接入手，在原文中寻求支持。也就是说，先对备选答案进行分析，确定它们之间的异同之处以及关键词，再在原文中寻找含有这些关键词的句子，然后仔细阅读理解，逐一排除直至最后得出正确的结论。恰当地使用此种技巧既能有效地利用时间，又能避免原文理解中不必要的干扰，因为我们是在利用关键词来寻找关键的信息。

总之，正确的结论，必须既合乎逻辑，又基于短文、源自文意。有理有据的结论是建立在客观事实或论据的基础上，而不应受到自己好恶的左右。作结论时，一定要以短文提供的信息为依据，再辅之以自己的常识经验。

Quiz

Read the following passages, and choose the best choice to the question.

1

Personality is to a large extent inherent—A-type parents usually bring about A-type offspring. But the environment must also have a profound effect, since if competition is important to the parents; it is likely to become a major factor in the lives of their children.

One place where children soak up A characteristics is school, which is, by its very nature, a highly competitive institution. Too many schools adopt the "win at all costs" moral standard and measure their success by sporting achievements. The current passion for making children compete against their classmates or against the clock produces a two-layer system, in which competitive A types seem in some way better have dangerous consequences: remember that Pheidippides, the first marathon runner, dropped dead seconds after saying: "Rejoice, we conquer!" By far the worst form of competition in schools is the disproportionate emphasis on examinations. It is a rare school that allows pupils to concentrate on those things they do well. The merits of competition by examination are somewhat questionable, but competition in the certain knowledge of failure is positively harmful. Obviously, it is neither practical nor desirable that all A youngsters change into B's. The world needs types, and schools have an important duty to try to fit a child's personality to his possible future employment. It is top management.

If the preoccupation of schools with academic work was lessened, more time might be spent teaching children surer values. Perhaps selection for the caring professions, especially medicine, could be made less by good grades in chemistry and more by such considerations as sensitivity and sympathy. It is surely a mistake to choose our doctors exclusively from A type stock. B's are important and should be encouraged.

(1) We can conclude that A-type individuals are usually _____.

    A) impatient   B) considerate

    C) aggressive   D) agreeable

(2) From the passage we can draw the conclusion that _____.

    A) the personality of a child is well established at birth

    B) family influence dominates the shaping of one's characteristics

    C) the development of one's personality is due to multiple factors

    D) B-type characteristics can find no place in a competitive society

(3) According to the last paragraph, the selection of medical professionals is currently based on _____.

A) candidates' sensitivity B) academic achievements
C) competitive spirit D) surer values

2

Where do pesticides fit into the picture of environmental disease? We have seen that they now pollute soil, water, and food, that they have the power to make our streams fishless and our gardens and woodlands silent and bird-less. Man, however much he may like to pretend the contrary, is part of nature. Can he escape a pollution that is now thoroughly distributed throughout our world? We know that even single exposures to these chemicals, if the amount is large enough, can cause extremely severe poisoning. But this is not the major problem. The sudden illness or death of farmers, farm workers, and others exposed to sufficient quantities of pesticides is very sad and should not occur. For the population as a whole, we must be more concerned with the delayed effects of absorbing small amounts of the pesticides that invisibly pollute our world.

Responsible public health officials have pointed out that the biological effects are cumulative over long periods of time, and that the danger to the individual may depend on the sum of exposures received throughout his lifetime. For these very reasons the danger is easily ignored. It is human nature to shake off what may seem to us a threat of future disaster. "Men are naturally most impressed by diseases which have obvious signs," says a wise physician, Dr. Rene Dubos, "yet some of their worst enemies slowly approach them unnoticed."

(1) According to the first paragraph, we can conclude that _____.

　　A) we should pay attention to the delayed effects of absorbing small amounts of the pesticides that invisibly pollute our world

　　B) human has already been polluted

　　C) pesticides result in the fishless in the rivers

　　D) pesticides are thoroughly distributed in our world

(2) The author believes that _____.

　　A) pesticides not only cause danger now but also leave threat for the future

　　B) people like to pretend to be part of the nature

　　C) attention should be paid to the delayed effects of pesticides pollution

　　D) those who are exposed to pesticides will die someday in the future

(3) It can be concluded from Dr. Dubos' remarks that _____.

　　A) people find invisible diseases difficult to deal with

　　B) attacks by hidden enemies tend to be fatal

　　C) people tend to overlook hidden dangers caused by pesticides

　　D) diseases with obvious signs are easy to cure

Part B   Reading

Technology's Ubiquitous Reach

Technology is rapidly changing our world. It seems that with each year the pace of change quickens. Each new process or invention makes still other advances possible. Such 19th and 20th century inventions as the telephone, the phonograph, the wireless radio, the motion picture, the automobile, and the airplane served only to add to the nearly universal respect that society in general felt for technology.

Human development is often divided into three categories: biological evolution, which gave us our brains; cultural evolution, which gave us the time to use them beyond survival; and technological evolution, arguably more responsible than any other process now in shaping what we are and what we're becoming. There is scarcely a domain of life that has not been radically transformed for the modern age by technology. It has changed the way we:

Work

Those burning eyes from endless hours at the computer all began 50 years ago with the development of the transistor, arguably the single most important technological breakthrough of the 20th century. Transistors gave rise to solid-state electronics, the integrated circuit, the personal computer, the Internet and, ultimately, the information age itself.

Transistors have done it all with only two basic functions: switching current off and on and amplifying it. Amplifying current led to the transistor radio and a host of other hand-held devices. Switching off and on makes the 0s and 1s that comprise the binary language computers use to analyze and communicate data.

Where we're headed: Transistors will continue to shrink so billions fit on a postage stamp-sized chip, creating smaller, faster, cheaper computers, cell phones and other gadgets.

Fight

No technological defined 20th-century warfare more than the atom bomb and radar, both of which have created unforeseen spinoff industries.

The atom bomb ended World War II and quietly fought the Cold War for four decades. Building the bomb was a technological feat unlike any other in its day. Creating the bomb from theory required a secret effort that rivaled the auto industry of its day. For all its apocalyptic ferocity, it may have put an end to perhaps the greatest plague of the 20th century: world war.

RADAR stands for Radio Detection and Ranging, a technology invented largely by the British, who noticed radio signals bounced off large objects like ships. Historians argue radar played a greater role in winning WWII than the bomb, giving British pilots early warning German bombs were crossing the English Channel, while allowing pilots to bomb through clouds and ships to track the Nazi U-boats.

Today, it's used to track hurricanes, guide passenger jets and peer into deep space at objects that elude our optical telescopes.

Where we're headed: Ever-smarter weapons such as robotic tanks and drone planes. To counter the growing threat of biological warfare, scientists are developing new effective against a broad range of microbes and other countermeasures aimed at thwarting bio-warfare organisms by scrambling their genes.

Heal

The invention of antibiotics transformed modern medicine because it finally gave physicians a way to intervene in infectious disease. Alexander Flemming discovered bacteria-killing mold called penicillin in 1928, but it wasn't until 1942 that scientists showed it could be used to treat infections.

Unit Eleven   Scientific Discoveries and Inventions

These wonder drugs, however, have also created new strains of bacteria that resist them. The 20th century also gave rise to the dread of every school child: the vaccine.

Where we're headed: Gene therapy and tissue replacement. Long overhyped in the 1990s, gene therapy could eventually provide cures for some of the most complex and stubborn diseases, those caused by defective genes.

Recent success in growing human embryonic cells (cells which have not yet decided to be heart cells or kidney cells and theoretically could be coaxed into becoming either) opens the possibility of growing replacement organs.

Travel

Technology gave us altitude and speed. We came into the 20th century behind the wheel of the Model T and left on the wings of the jumbo jet. The Boeing 747 can carry more people farther than any other passenger jet, explaining why it has dominated long-distance air travel for almost 30 years. Its economy class cabin is 30 feet longer than the Wright brothers' first flight at Kitty Hawk.

Where we're headed: Larger, faster and lighter jumbo jets capable of flying themselves under the guidance of satellite systems.

Explore

$E=mc^2$ may be the most famous equation of the century. But another—$V=c\ loge\ mi/mf$—opened the heavens. That's the formula that explains how much thrust a rocket needs based on its weight and other characteristics.

It paved the way for NASA's early Mercury missions, the conquest of the moon and now the recent development of the international space station.

Where we're headed: NASA officials insist we're going to Mars in 20 to 30 years, but others claim we won't get there for at least 50. Until then, we'll be mass-mailing cheap, robotic probes across the solar system.

Communicate

Early this century, phone service was a luxury. Now the same could be said of high-bandwidth (superfast) Internet connections.

The Internet evolved largely out of research and defense communications. Its "packet-switching" data transmission system grew out of the military's prestigious Defense Advanced Research Projects Agency, which had been looking for a way to share research information among its scientists.

In 1980, an information technologist at the European laboratory for nuclear studies in Switzerland created a program called "Enquire Within". It allowed users to access data "pages" with random associations. Ten years later, the idea was revived and enhanced as the World Wide Web.

Where we're headed: Wider Internet bandwidth to the home from a variety of vendors, including phone companies and cable TV providers, and cheap long-distance calls via the Internet. This is already happening. How it will play out is anyone's guess.

Live at home

Television, air conditioning and plastics—three inventions that are today still both praised and despised—have dramatically altered the landscape of the living room.

The first commercial television show aired, appropriately enough, at the 1939 World's Fair, an event billed as the beginning of the future. A decade later, there were 7 million sets in America. In 1951 the first color sets went on sale.

The first plastics were made as replacement for the ivory used in billiard balls and dental plates. But since 1976, plastic has become the nation's most widely used material.

Engineer Willis Carrier invented the first air conditioner in 1902 to cool the workshops at a Brooklyn publishing company where humidity was causing prints to bleed their colors. It wasn't until the 1920s that it caught on with the public as people escaped the heat in the newly air conditioned movie houses. It has since paved the way for the development of the new South and the exploding desert culture of the Southwest.

Where we're headed: High-definition TV, plastics that conduct electricity and "cool cities", designed to minimize the intense heat that builds up on concrete and blacktop.

Reproduce

The 20th century has seen the rise of technology to prevent conception among the fertile and fertilize the infertile. In 1954, Boston gynecologist Dr. John Rock first tested a pill that contained the hormone progesterone, hoping it would fool the system into thinking it was pregnant and stop ovulation. The results: 50 volunteers, not one ovulation. The "Pill" was born. Twenty-four years later, the first test-tube baby, Louise, was born in England. The slightly premature birth was the culmination of a procedure that began when an egg was removed from one of the mother's ovaries, fertilized with the father's sperm in a petri dish and then inserted in the mother's uterus through a tube.

Where we're headed: Cloning. Success with sheep, cows and mice lays the groundwork for cloned human. Like it or not, someone is probably going to do it.

New Words and Expressions

ubiquitous 无处不在的
phonograph 留声机，唱机
domain（活动、思想等的）领域，范围
solid-state [物] 固态
amplify 放大，增强，扩大
gadget 精巧的装置；小玩意儿
spinoff 副产品，派生物
rival 与……竞争；与……匹敌，比得上
apocalyptic 预示灾祸的，预示世界末日的
ferocity 凶残，残忍；残暴的行为
plague 瘟疫，灾害

bounce off（无线电信号等）跳动，反跳
elude 逃避，躲避
U-boat 德国潜艇
effective 有生力量
microbe 微生物，细菌
countermeasure 对策；干预措施
bio-warfare 生物战，细菌战
intervene (in) 干涉，干预；调停
strain（菌）株，（菌）系
thwart 反对，阻挠
dread 畏惧，恐惧
overhype 过分渲染
defective 有缺陷的，有毛病的
embryonic 胚芽的；胎儿的
coax 用好话劝，哄诱，劝诱
bandwidth 带宽
vendor 卖主

despise 鄙视，藐视
bleed（染料、油漆等）褪色，渗色，渗开
blacktop 沥青质原料，柏油路
fertile 能繁殖的
infertile 不孕的
gynecologist 妇科医生
hormone 荷尔蒙
progesterone 孕酮；黄体激素
ovulation 排卵，产卵
Pill 避孕药
culmination 顶点，高潮的到达
ovary 卵巢
sperm 精子
petri dish 有盖培养皿
uterus 子宫
groundwork 基础；底子，根基

Exercises

I. Find out the English equivalents to the following Chinese terms from the passage.

1. 晶体管
2. 集成电路
3. 光学望远镜
4. 无人驾驶飞机
5. 青霉素
6. 疫苗
7. 基因治疗
8. 组织放置
9. 胚胎细胞
10. 大型喷气机
11. 高带宽因特网接入
12. 世界博览会

Unit Eleven   Scientific Discoveries and Inventions

13. 台球
14. 酷热
15. 荷尔蒙孕酮
16. 试管婴儿

II. **Answer the following questions by using the sentences in the text or in your own words.**

1. In what ways have technologies changed our life?
2. What are the three categories of human development?
3. What is the single most important technological breakthrough of the 20th century?
4. What are the most influential technologies that have changed the 20th century warfare?
5. What is radar currently used for?
6. What is the most important invention in medicine in the 20th century?
7. What are the two directions that technology would affect our health?
8. Which formula opened the heavens and paved the way for the development of the international space station, $E=mc^2$ or $V=c\ log_e\ mi/mf$?
9. What are the three objects that have changed the landscape of our living rooms?
10. What do you think scientists will do after they have cloned sheep, cows and mice successfully?

Part C   Extended Reading

The Telephone and Its Inventor

In the nineteenth century, the invention of the telephone made it possible to send noises, signals, and even music over wires from one place to another. However, the human voice had never traveled this way. Many inventors tried to find a way to send a voice over wires, and in 1876 some of their efforts were crowned with success. Two American inventors, Alexander Graham Bell [1] and Elisha Gray [2], succeeded at almost the same time. The United States Supreme Court finally had to decide which of the two was the first inventor of the telephone. The Court decided in Bell's favor.

Born in Edinburgh, Scotland, Bell grew up in a family that was very interested in teaching people to speak. His grandfather had been an actor who left the theatre to teach elocution; his father was a teacher who helped deaf-mutes learn how to speak.

After studying in Edinburgh and London, Bell moved to Canada with his family. Bell was frail and, since two of his brothers had died of tuberculosis, his family felt that the climate of Canada would be less damp and healthier for Alexander.

In 1871, the year after the Bell's arrival in Canada, young Alexander found a job in the United States. He was hired as a teacher in a new school for the deaf in Boston, Massachusetts. He taught there and at other United States schools. Eventually, he opened a school of his own. One of his pupils became his wife.

Bell thought that if he could make speech visible by actually showing the vibrations of the voice, he could teach his deaf pupils to make the same vibrations themselves. Bell tried to find a way to reproduce the vibrations by

Unit Eleven Scientific Discoveries and Inventions

electrical means. He was led to believe that if a wire could carry vibrations, there was no reason why it could not carry the words spoken by a person. After all, he reasoned, spoken words are only a series of different vibrations.

Bell spent his evenings experimenting with tuning forks (fork-shaped pieces of metal which always produce the same musical note and which are used to tune pianos), metal springs, and magneto batteries. His assistant, named Watson, worked with him for almost four years. They spent all their spare time, as well as their spare money, on their experiments. When their money ran out, they persuaded some businessmen to help them financially.

Finally, one day in 1874, Bell was at his receiver in one room, trying to catch the signals sent by Watson from another room. Suddenly he heard a faint twang. Rushing into Watson's room, Bell shouted, "What did you do then?" He found that the sound had been made accidentally when two pieces of metal had stuck together and closed the circuit. Watson had pushed the pieces apart, and the metallic vibration had been carried to Bell over the wire. After this discovery, all they had to do was find the best kinds of materials for carrying the human voice over wires. The basic principle had been discovered.

After much experimentation, the first telephone was exhibited at the Centennial Exposition of 1876 in Philadelphia. At first few people paid any attention to the young man whose table held a curious box with an unexciting appearance. There were many other things to see at the exposition, and Bell was not the "barker" type who painted big signs and shouted about the wonders of his display. However, when the judges came to inspect the new machine, they were amazed. Some of the visitors at the exposition included the English scientist, William Thomson [3] (Lord Kevin), and the Emperor of Brazil, who had previously met Bell. Thomson said the telephone was "the greatest marvel hitherto achieved by the electric telegraph". The Emperor of Brazil Exclaimed, "It talks!" Soon crowds were gathering around the exhibit and, from that time on, Bell was famous.

Bell never lost his interest in helping the deaf. After all, it was because of his work with them that he had begun his research for a way to reproduce the human voice electrically. If he had not tried to make speech visible to the deaf, he might never have discovered the telephone. Bell established a fund for the study of deafness. His studies showed that deafness was increasing in America, because deaf people were kept in institutions where they met and married one another. This policy, he warned, was producing a deaf variety of the human race. He strongly advocated a policy where deaf people could be taught to use language instead of being limited to the use of sign language. Bell said that the deaf should also live among normal people, so that they would not consider themselves inferior to others and fail to develop their powers of speech.

Not long after Bell's invention, telephone companies were established in the United States, Great Britain, France and many other countries. It is said that in Bell's later life, when telephone disturbed him because it was always ringing and interrupting his experiments.

Bell invented other things besides the telephone, but none of them were as important to mankind as the telephone. One of the most interesting of his inventions was the photophone, which carried the sound of the voice for a short distance over a vibrating beam of light. Another invention of interest was the graphophone, an ancestor of the phonograph, which used wax records similar to those we know today.

However, probably none of these later inventions gave Bell the same feeling of triumph as he had on the day when he spilled some acid from his batteries. It was after he had worked for months to find ways to send something more than metallic twangs over the wires. Thinking Watson, his helper, was in the next room, Bell called, "Mr. Watson, come here. I want you." Watson was not in the next room. He was down in his lab, next to the receiver. To Watson's surprise, he heard the words perfectly. He ran to tell Bell the news: the wires had carried Bell's voice perfectly.

Unit Eleven　Scientific Discoveries and Inventions

New Words and Expressions

elocution 演说术；朗诵法
deaf-mute 聋哑者
frail 虚弱的；脆弱的
vibration 颤动，振动，震动
tuning 调音，定弦；起音，定音
magneto 磁石发电机
receiver 电话听筒，受话器
twang 拨弦声；"砰"的一声
exposition 展览会，博览会
barker 大声招徕顾客者
hitherto 迄今，到目前为止
graphophone 格拉福风留声机，留声机
wax 蜡剂

Notes

1. Alexander Graham Bell（1847—1922）亚历山大·格拉汉姆·贝尔，美国发明家和企业家，被世界誉为"电话之父"。他获得了世界上第一台可用的电话机的专利权，创建了贝尔电话公司（AT&T 公司的前身）。

2. Elisha Gray（1835—1901）埃利萨·格雷。电话是贝尔和格雷在不同背景下同时取得的一项相同的发明，但为什么社会只公认贝尔是电话的第一发明人呢？这是因为格雷虽然和贝尔同年发明了电话，但由于格雷申请电话发明专利比贝尔晚了 2 个小时，所以也只能榜上无名。

3. William Thomson（Lord Kevin）（1824—1907）威廉·汤姆森（开尔文勋爵）是一位在北爱尔兰出生的英国数学物理学家、工程师，也是热力学温标（绝对温标）的发明人，被称为"热力学之父"。

Exercises

Answer the following questions by using the sentences in the text or in your own words.

1. What was the interest of Bell's family?
2. Why did Bell's family move to Canada?
3. What did Bell do in 1871?

4. Where was the first telephone exhibited?
5. How did people react to his first telephone at first?
6. How did people react to his first telephone when the judges came to inspect this new machine?
7. Besides telephone, what else did Bell invent?

Unit Twelve

Great Men and Women

Part A Lecture

判定作者的观点、态度和语气

一般说来，作者的观点、态度和感情就是一篇文章的基调。在阅读中为了获取作者传达的全部信息，就必须了解作者的思想及感情倾向。有时，作者会直接发表见解，态度明朗；而更多的时候，作者却把见解寓于字里行间，态度含蓄。表述得比较曲折委婉的文章就需要我们具备识别作者写作观点和褒贬态度的能力。如果我们在阅读中把握不住作者的态度和文章的语气，就意味着我们对文章内容以外的许多东西还没有完全理解，对文章内容及主题的理解只能说还是比较肤浅的。作者的态度与语气建立在对主题的理解基础之上，又超越了主题。因此，通过阅读判定作者的观点、态度和文章的语气是体现阅读能力的重要方面。

1 判定作者观点的技巧

1·1 对作者观点的理解

文章离不开事实和观点。在阅读时，我们首先应能分辨事实和观点。事实是指已经发生的或所做之事，因而在复述事件的时候，一般都用过去时。所以我们可以从时态上找到一些线索。同时事实还指众所周知被大家接受的东西，比如属于真理的一类的东西，"太阳从东边升起"（The sun rises in the east）等，这类事实的叙述用一般现在时。另外，事实还指现实、真实的事物、存在的事物。如：

Unit Twelve　Great Men and Women

A. China, India, the Soviet Union, and the United states have a total population of 1,790,000,000.

B. She became the only black woman president in the country.

分析：由于观点是表达个人的看法，我们可以通过下面一些表达观点的线索做出判断。

1）通常表示对某事物进行判断和评论的形容词，如：ugly、pretty、safe、dangerous、clever、stupid、well-dressed、sloppy、desirable、hateful；

2）表示逻辑可能的情态动词，如：can/could、may/might、will/would、tend to、should；

3）清楚表明并将作者观点引出的词和短语，如：believe、think、feel、suggest、in my opinion、in my view、according to me、claim、know、suppose、think、seem、consider、agree、contend；

4）一些表达可能性或不完全肯定的词，如：probably、perhaps、likely、maybe、possible、plausible、usually、often、sometimes、on occasion；

5）间接表示作者态度的小句，如：It is likely that、It seems（appears）that、It seems/appears that...、It's obvious/clear that、It's no secret that、It's no longer necessary to、It is indispensable、...are still a must、There is a need for、...have the right、It is essential for、shouldn't、can't have to 等。

A. "It takes too **long** to watch a baseball game."

B. "You are a **reckless** driver."

分析：上例中的形容词 long 和 reckless 明确表达了作者的观点。

分析下列句子，辨认哪些是事实，哪些是观点。

A. Portsmouth is the best place for a navy base.

B. Television commercials are boring.

C. The school cafeteria has great food.

D. Ottawa is the capital of Canada.

E. Alfred Nobel was born in Stockholm on October 21, 1833.

F. The first public protests against the low status of women were made in the 1830s.

G. Water boils at 100℃.

分析：观点往往体现一个人的判断能力，他对某人或某事物是怎么看的。因此，在表示观点的陈述中，常常带有修饰词，如上列句 A、句 B、句 C 中的"best""boring""great"。而事实则是能够被证实的，实际上是存在的或被人们普遍承认的东西，如上列句 D、句 E、句 F、句 G。

我们知道，作者在写文章时，常常用具体的事实来阐明自己对人或事物的见解和看法。这些见解和看法，就是作者的观点。

1.2 推断作者观点的技巧

一篇文章不可避免地反映了作者的观点，只要我们阅读时仔细琢磨，抓住中心思想，注意文中用词造句的特点以及段落展开模式，那么我们就会发现作者的观点还是有迹可循的。具体来讲，可以从以下四个方面推断作者的观点。

▶ **1.2.1 通读全文，理解中心思想和主要事实，推断作者的观点。**

Money spent on advertising is money spent as well as any I know of. It serves directly to assist a rapid distribution of goods at reasonable prices, thereby establishing a firm home market and so making it possible to provide for export at competitive. By drawing attention to new ideas, it helps enormously to raise standards of living. By helping to increase demand, it ensures an increased need for labor. And it is therefore an effective way to fight unemployment. It lowers the costs of many services: without advertisements your daily newspaper would cost four times as much, the price of your television license would need to be doubled and

Unit Twelve Great Men and Women

travel by bus or tube would cost 20 percent more.

And perhaps most important of all, advertising provides a guarantee of reasonable value in the products and services you buy. Apart from the fact that twenty-seven Acts of Parliament govern the terms of advertising, no regular advertiser dare promote a product that fails to live up to the promise of his advertisements. He might fool some people for a little while through misleading advertising. He will not do so for long, for mercifully the public has the good sense not to buy the inferior article more than once. If you see an article consistently advertised, it is the surest proof I know that the article does what is claimed for it, and that it represents good value.

Advertising does more for the material benefit of the community than any other force I can think of.

There is one more point I feel I ought to touch on. Recently I heard a well-known television personality declare that he was against advertising because it persuades rather than informs. He was drawing excessively fine distinctions. Of course advertising seeks to persuade.

If its message were confined merely to information and that in itself would be difficult if not impossible to achieve, for even a detail such as the choice of the color of a shirt is subtly persuasive—advertising would be so boring that no one would pay any attention. But perhaps that is what the well-known television personality wants.

(1) In the author's opinion, _____.
   (A) advertising can seldom bring material benefit to a man by providing material
   B) advertising informs people of new ideas rather than wins them over
   C) there is nothing wrong with advertising in persuading the buyer
   D) the buyer is not interested in getting information from an advertisement

(2) The author deems that the well-known TV personality is _____.
   A) very precise in passing his judgment on advertising
   B) interested in nothing but the buyers' attention
   C) correct in telling the difference between persuasion and information
   D) obviously partial in his views on advertising

分析：纵观全文，我们可以推断 C 项是问题 1 的正确答案；因为在短文的前三段中，作者阐述了广告由于提供信息给人们带来的物质利益。从第三段所谈到的"Advertising does more for the material benefit of the community than any other force I can think of"，我们可以排除 A 项。通过作者在第四段反驳著名电视人物的观点时，说到的"of course advertising seeks to persuade"，以及在最后一段表明的"persuasive"在广告中的重要作用，我们可以排除 B 项。而 D 项与短文内容不符。

从短文中可以找出作者对电视人物的讽刺性措辞"He was drawing excessively fine distinctions"和文中的"of course advertising seeks to persuade"，以及最后一段作者强调的 persuasive 在广告中的重要性，我们可以推断出题 2 选项 D 为正确答案。

▶ 1.2.2 注意作者观点的隐含表达，如委婉表达、反义表达等。

An Eskimo baby who was brought up by American parents would speak English, hate castor oil, and act like any other American child; and an American baby who was brought up by an Eskimo family would grow up to be a seal hunter, to like eating blubber, and to speak Eskimo.

分析：短文中作者没有直接表达自己的观点，而是把观点隐含在字里行间。读了这段短文，我们可以推断作者的观点为：Different environments make different people.

▶ 1.2.3 注意分清哪些是作者引用别人的观点，哪些是作者自己的观点。

Some people believe that international sport creates goodwill between the nations and that if countries play games together they will learn to live together. Others say that the opposite is true: that international contests encourage false national pride and lead to misunderstanding and hatred. There is probably some truth in both arguments, but in recent years the Olympic Games have done little to support the view that sports encourages international brotherhood. Not only was there the tragic

Unit Twelve Great Men and Women

incident involving the murder of athletes, but the Games were also ruined by lesser incidents caused principally by minor national contests.

According to the author, recent Olympic Games have _____.

A) created goodwill between the nations

B) bred only false pride

C) barely showed any international friendship

D) led to more and more misunderstanding and hatred

分析：短文中作者首先用"Some people believe..." "Others say..."，引出了两种对立的观点。接着作者才提出自己的观点"There is probably some truth in both arguments, but in recent years the Olympic Games have done little to support the view that sports encourages international brotherhood."（两种观点也许都有些道理，但近年来奥运会没有做出什么事来支持运动，促进国际间兄弟情谊。）。接着作者还用事实来进一步表明自己的观点（文中最后一句话）。由此，我们得知 C 为正确选项。

▶ **1.2.4 注意文章中出现的"yet" "however" "but" 等转折词。**

有时作者在这些转折词后面表明自己的观点。要正确判断作者的观点，必须将上下文联系起来。

This is not just a sad-but-true story; the boy's experience is horrible and damaging, yet a sense of love shines through every word.

How does the author of this sentence feel about the story?

A) It transmits a sense of love.

B) It is just sad.

C) It is not true.

D) It is horrible and damaging.

分析：转折词"yet"后面的意思是作者真正想表达的观点，故 A 项为正确答案。

2 推断作者态度的技巧

2.1 对作者态度的理解

在文章中，作者通过直接或间接的语言描写或修饰手段来表达自己对作品中人、事、物的态度。如：

The shops of the border town are filled with many souvenirs（纪念品）, piñatas（彩饰陶罐）, pottery, bullhorns and serapes（用作披肩的彩色毛毯）, all made from cheap material and decorated in a gaudy（华丽而俗气的）manner, which the tourist thinks is true Mexico folk art.

The attitude of the author of this sentence toward the souvenirs can best be described as _____.

A) positive   B) indifferent   C) admiring   D) contemptuous

分析：题中四选项列出了四种不同的情感态度。根据作者的遣词造句如修饰语 cheap 和 gaudy 以及带点儿讥讽意味的"true"一词，我们推断作者对 souvenirs 的态度是轻蔑的。故 D 项表明作者的态度。

2.2 推断作者态度的技巧

在推断作者的态度时，我们应注意以下几点：

2.2.1 作者使用的描写性段落和描写性语言

Every afternoon a line of very old women passes down the road outside my home, each carrying a load of firewood. All of them are tiny. It seems to be generally the case in primitive countries that the women, at a

Unit Twelve Great Men and Women

certain age, shrink to the size of children. One day a poor old creature who could not have been more than four feet tall crept past me under a vast load of wood. I stopped her and put a five-sou piece (less than a penny) into her hand. She answered with a shrill wail, almost a scream, which was partly gratitude but mainly surprise. I suppose that from her point of view I seemed almost to be violating a law of nature. She accepted her status as an old woman, that is to say, a beast of burden.

The author's attitude toward his subject is _____.

    A) resentful   B) distrustful   C) sympathetic   D) affectionate

分析：短文中作者首先描写了一个画面：每天下午一队肩负木柴重担的年迈妇女从门外的路上走过，她们都非常矮小。作者又描述了作者与其中一位老人相遇的情景：一天作者遇到一位可能还不到四英尺高的可怜老人，背着一大担木柴从作者面前几乎是爬行而过。作者叫住了她并塞给她一枚硬币。从作者描述这些老妇人所采用的白描手法、对比手法（could not have been more than four feet tall 与 under a vast load of wood）、修饰词（poor old creature）以及动作描写（crept），还有作者塞给其中一位老人一枚五分硬币的事实，我们不难看出，作者对自己描写的人物的态度是同情的。故 C 为正确答案。

▶ 2.2.2 作者在文章中所用的直接表达感情和态度的词语

    I disagreed then as now with many of John Smith's judgments, but always respected him and this book is a welcome reminder of his big, honest, friendly, stubborn personality.

    A) He dislikes him but agrees with his idea.

    B) He considers him to be a disagreeable person.

    C) He disagrees with his ideas but respects him.

    D) He disagrees with him then but agrees with him now.

分析：根据句中所用的直接表明态度的词"disagreed"和"respect"，我们知道选项 C 为正确答案。

2.2.3 作者在论题中使用对比的论据及阐述的重点倾向

We will not deny that tourism has an influence which is not always appreciated by everybody. Some people prefer to be left alone. They are not interested in the money which tourists spend in shops and hotels, as well as on taxis and coach tours. But the majority of ordinary people have far more to gain from tourism than they have to lose.

What is the writer's attitude toward tourism?
  A) He is against tourism.
  B) He is not at all interested in tourism.
  C) He is in favor of tourism.
  D) We can't tell from the passage.

分析：文中用了两个论据：1）有些人更愿意不被打扰。他们对旅游者花在商店里、旅馆里、出租车上和乘车旅游上的钱都没有兴趣。2）大多数普通人要得到的比要失去的多得多。这两个论据对比，作者的观点倾向于后者。因为一般表达重点都在 but、yet、however 之后。此外，我们还可将 some 与 majority 对比，have far more to gain than they have to lose 对比。据此，我们可推断作者对旅游业的态度是赞成的。故选项 C 为正确答案。

2.2.4 文章的主题思想及作者对题材的选择

Nursing at Beth Israel Hospital produces the best patient care possible. If we are to solve the nursing shortage, hospital administration and doctors everywhere would do well to follow Beth Israel's example.

As Beth Israel each patient is assigned to a primary nurse who visits at length with the patient and constructs a full-scale health account that covers everything from his medical history to his emotional state. Then she writes a care plan centered on the patient's illness but which also includes everything else that is necessary.

Unit Twelve   Great Men and Women

The primary nurse stays with the patient through his hospitalization, keeping track with his progress and seeking further advice from his doctor. If a patient at Beth Israel is not responding to treatment, it is not uncommon for his nurse to propose another approach to his doctor. What the doctor at Beth Israel has in the primary nurse is a true colleague.

Nursing at Beth Israel also involves a decentralized（分散的）nursing admini-stration; every floor, every unit is a self-contained organization. There are nurse-managers instead of head nurses; in addition to their medical duties they do all their own hiring and dismissing, employee advising, and they make salary recommendations. Each unit's nurses decide among themselves who will work what shifts and when.

Beth Israel's nurse-in-chief ranks as an equal with other vice presidents of the hospital. She also is a member of the Medical Executive Committee, which in most hospitals includes only doctors.

The author's attitude towards the nursing system at Israel Hospital is _____.
   A）neutral   B）positive   C）critical   D）negative

分析：作者在文章的第一段（Introductory paragraph）中提出了自己要表达的主题思想。注意文中的措辞"best""would do well to follow Israel's example"。接着作者选择了 Beth Israel 独特和细致周到的护理、医生与护士的密切配合、独特的护士管理方式、护士的地位等具体事实含蓄地与其他医院进行对比，突出了文章的主题思想。

根据作者的选材、措辞及文章的中心思想，我们可推断作者对 Israel's nursing system 的态度是肯定的，故选项 B 为正确答案。

3 判断作者语气或笔调的技巧

3·1 对作者的语气、笔调的理解

作者的语气、笔调是作者的观点、态度的一种表现形式。作者写文章时要对某一个问题表达自己的立场、观点、态度或看法；但不直接表现，而是

在用词造句的种种方式中流露出来，这种流露表明赞成或反对，表扬或批评，公正或偏执等情绪。因此，对作者语气、笔调的理解，是从另一个角度考核我们对作者思想观点倾向及态度的理解。如：

  Turn on the world news broadcast any evening, and the predominant mood is one of semi-darkness and hopelessness. Maybe Brazil and Peru haven't gone to war, but the news is that some other countries have. Thousands of people have been left homeless by earthquakes, floods, and fires, but nobody reports on the millions of people unharmed by natural disasters. In the cities, men and women go about the daily affairs of earning a living, quietly and calmly, without making the news, but crime, greed, and corruption seem to be on every street corner according to the latest news report.

The tone of the passage would be _____.
  A) pessimistic   B) angry   C) one of gloom   D) critical

分析：这是一篇议论文，议论的对象是世界新闻广播。作者在段落开始的主题句中指出："...and the predominant mood is one of semi-darkness and hopelessness."即"……（新闻广播）的基调是阴晦的和绝望的"，这便是短文议论的主题。作者定调之后，用后续的三个 supporting sentences 作为事例来揭示这种基调，而且每句都用转折性的逻辑承接词 but 作正反面对比。这种对比阐述方法加强了议论的说服力。纵观全文，其用语是平和的，没有激烈言辞，没有悲观的流露，没有愤怒，也无沮丧，但可以领会到其中的不满——即批评的语气或笔调，故选 D。

3.2 判断作者的语气或笔调的技巧

  我们在判断作者的语气和笔调时，首先要通读全文，对文章的写作风格有个了解，尤其要仔细阅读文章主体部分，了解文章的主题思想，而这些通常能帮助判断作者的写作笔调。

  In only two decades Asian Americans have become the fastest-growing U.S. minority. As their children began moving up through the

Unit Twelve Great Men and Women

nation's schools, it became clear that a new class of academic achievers was emerging. Their achievements are reflected in the nation's best universities, where mathematics, science and engineering departments have taken on a decidedly Asian character. This special liking for mathematics and science is partly explained by the fact that Asian-American students who began their educations abroad arrived in the U.S with a solid grounding in mathematics but little or no knowledge of English. They are also influenced by the promise of a good job after college. Asians feel there will be less unfair treatment in areas like mathematics and science because they will be judged more objectively. And the return on the investment in education is more immediate in something like engineering than with an arts degree.

Most Asian-American students owe their success to the influence of parents who are determined that their children take full advantage of what the American educational system has to offer. An effective measure of parental attention is homework. Asian parents spend more time with their children than American parents do, and it helps. Many researchers also believe there is something in Asian culture that breeds success, such as ideals that stress family values and emphasize education.

Both explanations for academic success worry Asian Americans because of fears that they feed a typical racial image. Many can remember when Chinese, Japanese and Filipino immigrants were the victims of social isolation. Indeed, it was not until 1952 that laws were laid down giving all Asian immigrants the right to citizenship.

The author's tone in this passage is _____.

A) sympathetic   B) doubtful   C) critical   D) objective

分析：本文反映亚裔美国人尤其是其子女在学业方面的情况。作者指出亚裔美国学生偏爱理科，基础扎实，成绩出色。另外也指出这与家长的关心、家庭观念、重视教育等方面有关。从整体上作者比较客观地反映这一现实。因此，答案 D 能说明作者的态度、笔调。

其次，要注意作者在描述事物、表达观点时遣词造句的方式。因为在表达个人看法时，作者往往用一些带有个人感情色彩或褒贬分明的词汇，对这些多加注意就能做出正确的判断。

Incidents of this kind will continue as long as sport is played competitively rather than for the love of the game. The suggestion that athletes should compete as individuals, or in non-national teams, might be too much to hope for. But in the present organization of the Olympics there is far too much that encourages aggressive patriotism.

According to this passage, the author's mood is _____ to the organization of the Olympics.

A) positive　　　　B) negative　　　C) neutral　　　D) indifferent

分析：本题要求推断作者的叙述语气。短文中作者写道：只要体育运动是为竞争而不是因为热爱去进行，这类事件就会继续发生。有人建议运动员以个人身份参加，不以任何国家队队员身份参加比赛，这可能过于奢望。但目前的奥林匹克组织过多地助长了偏激的爱国主义。从短文中可看出作者是对其不满的，持的是一种批评的态度。特别是最后一句。因此，只能选 B。在本题中，对短文中某些关键词或句的理解尤显重要，如"But in the present organization of the Olympics there is far too much that encourages aggressive patriotism."等。

4 综合练习

从以上分析可知，就作者论述时的思想倾向和对论述对象所持的态度做出判断，即判断出作者对所陈述的观点（论述的对象）是赞同、反对还是犹疑不决，对记叙描写的人、事、物是赞颂、同情、关心还是冷漠、厌恶。作者的这种思想倾向和感情色彩往往隐含在字里行间，或流露于某些修饰性词语之中。因此，我们既要依靠短文的中心思想作为推理的前提，又要注意作者的措辞，尤其是形容词一类的修饰语。

在具体判定作者观点、态度和短文语气的过程中，我们可以从以下三个角度去捕捉信息，进行推测：

Unit Twelve Great Men and Women

（1）短文内容

　　A. 研究短文的段首和段尾句（主题句），确定短文主题；

　　B. 注意能揭示作者感情、态度、观点的词语；

　　C. 注意展露段落语气和作者态度的句子和句子结构特点；

　　D. 研究描写性段落以及文中人物对某一场面进行描述的语言。

（2）短文写作

　　A. 注意作者处理主题、展开段落的方式；

　　B. 确定作者劝说、解释、说明、逗趣、提供信息等写作意图。

（3）作者主观意向

　　A. 洞见作者对自己论述对象的态度；

　　B. 揣测作者对读者所持的态度；

　　C. 了解作者为争取影响读者可能运用的情感手段；

　　D. 判断整个段落或短文所形成的总气氛。

以下我们通过实例练习判定作者观点、态度和语气。

During the first week of September, 1900, everybody went swimming on Galveston Beach, Texas. There'd never been such a fine surf—great rolling combers that swept in from the Gulf. Yet there was hardly a breath of wind.

A blanket of humid heat lay over the city. Storm warnings had gone out of the Gulf shipping companies. The barometer was falling. Those signs should have been of concern to the people of a town built on a sand bar only nine feet above the sea at its highest point. But nobody seemed to be worried. Scientists had said that the city was safe from storm and flood because the long, gentle slope of the sea bottom would protect it.

On Friday afternoon of that week the sea-bathing had to be stopped. The surf was becoming too dangerous. Still there was no wind. The surface of the Gulf water was smooth, gray satin lined with streaks of foam. Older citizens began to study the sky toward the southeast, toward the Caribbean where hurricanes are born.

(1) The overall tone of the selection is one of _____.
   A) gloom, sadness   B) liveliness, gaiety
   C) uneasiness, danger   D) hopelessness, horror

(2) The general attitude of the Galveston residents during the first week of September 1900 suggests that they were _____.
   A) impatient and bored with the calm weather
   B) calm and unconcerned about the possible danger
   C) excited and expectant over the approaching storm
   D) tense and resentful over the inaccurate weather forecasts

(3) The selection ends on a note of _____.
   A) fear   B) hopefulness   C) worry   D) indifference

(4) In Paragraph 2, "Those signs should have been of concern to the people of a town built on a sand bar only nine feet above the sea at its highest point" indicates that the author views the residents with some _____.
   A) impatience   B) indifference   C) humor   D) disgust

(5) We can infer from the passage that _____.
   A) a surf can appear without wind
   B) the first week of September, 1900 was very hot on Galveston, Texas
   C) a hurricane was coming on Friday that week
   D) all of the above

分析：本短文主要叙述得克萨斯州 Galveston 海滨九月第一周气候的异常和预感到灾难性天气到来的不安。短文共 14 句，有 6 句为超短句，1 句并列短句，创造了一种飓风到来前紧张、担心的总气氛。短文第一段最后一句和第二段第 1、3 句两句的描述，透露出人们对异常天气的察觉和作者的担心；第二段倒数第 2 句、第三段第 2、3 句这三个短句，又暗示了一种什么危险的存在；第一段 Yet there was hardly a breath of wind 和第三段 Still there was no wind 两个短句意义上的重复，揭示了情况的紧迫，并暗示了可能的突变一定与 wind 有关；最后一句则进一步确定可能的不测是飓风，年长的居民和作者的担心都在这一句有所流露。整个短文的气氛是压抑、不安和紧迫。从第一段第 1、2 句和第二段第 4、5 句，又可推知 Galveston 的居民对异常天气情况和可能到来的恶劣气候持有的是粗心、过于坦然的无所谓态度，而对他们的这种态度，在字里行间，作者不无不满和责备。因为对可能的灾

Unit Twelve   Great Men and Women

难性天气变化，作者流露出的是担心和不安（of concern, worried）。

题1答案为C。从整篇短文看，占主导地位的气氛既没有"忧郁悲伤"（A项）和"绝望恐怖"（D项），也不是"轻松欢快"（B项），而应该是"不安和担心"（C项）；从第一段最后一句、第二段第4、5句和第三段第2、3句和最后一句可得到验证。

题2答案为B。由第一段第1、2句，第二段第4、5句可知，Galveston居民对天气可能会发生的突变是"不担心、无所谓"，而不是"激动和期待"（C项）。"平静的天气"（A项）和"气象预报不准确"（D项）则有悖文义。

题3答案为C。短文结尾讲到年长者开始注意并观察东南方飓风生成的加勒比海那个方向的天空，暗示人们不再是"漠不关心"（D项），而是因担心而注意观察起来了，但谈不上"害怕"（A项），自然更不是"乐观"（B项）。

题4答案为A。根据句中的情态动词和该镇只高出海平面九英尺的实际地理位置，不难推知作者对该镇居民的态度是有点不满因此是"着急"（A项），而不是"不在乎"（B项），但不至于"厌恶"（D项），当然更看不出有"幽默"（C项）。

题5是推论题，答案为D。因为A项、B项、C项都可由短文推知：A项由第一段和第三段第2、3句推知，B项由第二段第1句结合第一段第1句推知，C项由第二段第2句和短文最后一句推知。

  Fried foods have long been frowned upon. Nevertheless the skillet is about the handiest and most useful piece of kitchen equipment. Lumberjacks and others engaged in active work, who need 4000 calories or more a day, eat about one-third of their food fried. Meats, eggs, and French toast cooked in this manner are served in millions of homes daily. It seems that the people who eat these foods do not suffer more from indigestion than do those who insist upon broiling, roasting, or boiling. Some years ago one of our most famous doctors investigated the digestibility of fried potatoes. He found that the pan-fried variety was more easily broken down for digestion than French or deep fried potatoes. Even the deep fat variety, however, dissolved within the digestive tract more readily than boiled potatoes. Furthermore, he learned, by watching the progress of the contents of the stomach with X-rays, that fat actually increased the rate of digestion.

(1) The author implies that fried foods _____.
   A) may be harmful
   B) should be eaten daily
   C) may be eaten in moderate amounts
   D) should be avoided by inactive people

(2) The sentence, "Lumberjacks and others engaged in active work…eat about one-third of their food fried", suggests that _____.
   A) lumberjacks do not understand nutrition
   B) other methods of cooking are not available
   C) fried foods are hazardous to health
   D) fried foods supply energy

(3) Fried foods are popular because they are _____.
   A) economical   B) convenient
   C) good-tasting   D) nourishing

分析：

纵览全文，我们可以看出，作者对 fried foods 持肯定态度。从短文的开头、末尾以及 Lumberjacks… eat about one-third of their food fried 可以迅速排除题 1 中的 A 项和 D 项。另外，短文也没有说我们每天都必须食用 fried foods，而是以商量的口气建议人们适当加以食用。故答案为 C。

题 2 要求找出因果关系。"果"是他们每天所需食物中的三分之一是 fried foods，那么原因是 they engaged in active work and need 4000 calories or more，顺而推之，fried foods 能够提供其所需的热卡。故答案为 D。

题 3 答案为 B。短文第 2 句 Nevertheless the skillet is about the handiest and…，作者说 skillet 是厨房中最为方便的烹调用具，实际上 skillet 是用来 fry foods，其含义是 fried food is the handiest。

从以上实例分析可知，要准确判定作者的观点、态度和语气并非是件容易之事，需要我们把握文章的主题思想，还要注意文章中作者的措辞；同时，还需要我们多读多练，在实践中掌握这项重要的阅读技巧。

Quiz

Read the following passages and choose the best answer to each of the following questions.

1

Figures can be deceiving. For example, *Time* magazine recently reported that the average Yale graduate of the class of 1944 was making $35,111 a year. Well, good for him! But what exactly does that figure mean? Is it a proof that if you send your child to Yale you won't have to work in your old age and neither will he? What kind of sample is it based on? You could put one Texas oilman with two hundred hungry writers and report their average income as $35,111 a year. The figure is exact, but it has no meaning. In ways similar to this, the facts and figures pour forth every day. They are used to point out the truth, when in fact they inflate, confuse, and over-simplify the truth. The result is "number nonsense".

(1) The author's comment, "Well, good for him", as a response to the average yearly salary of a Yale graduate is meant to show _____.

A) humor B) praise
C) displeasure D) indifference

(2) In this selection the author uses tone to _____.

A) be scientifically objective B) create sympathy for writers
C) condemn Yale graduates D) reject the theory of average

(3) The tone in this passage can best be described as _____.

A) scientific and objective
B) sentimental and moving
C) concerned and informative

D) impersonal and matter-of-fact

2

It is an old joke that Americans are soon going to lose the use of their legs... but it is true that few Americans will walk anywhere if they can help it, either for practical purpose or for pleasure. You can do your banking from your car, without leaving the driving seat. You can mail your letters in postboxes that reach the level of your car window. You can watch a film from your car in a drive-in theater. At many stores you can be served in your car. At countless restaurants waitresses will hitch trays to the car door, so that you can eat without moving. In Florida there is even a drive-in church. And in California a funeral home has drive-in service for people who wish to purchase grave sites and caskets ahead of time.

(1) The tone of the paragraph is _____.

    A) positive   B) exciting

    C) hopeful   D) disapproving

(2) The author has created an atmosphere to illustrate _____.

    A) his sentimental and nostalgic mood

    B) the high standards of American life

    C) the style of life many Americans prefer

    D) the inventiveness of the American people

(3) Apparently the author feels that drive-in churches and funeral homes are _____.

    A) good for the economy   B) absurd and gross ideas

    C) wonderful conveniences   D) unpatriotic institutions

Part B Reading

Three Great Physicists

The development of modern science and its increasing impact on our daily life is one of the great stories of our time. Contributions to the development have come from dozens of figures from the history of science. They are great, successful and have achieved a great deal in their scientific research. This passage describes three great men in physics.

Galileo Galilei [1]

Galileo, perhaps more than any other single person, was responsible for the birth of modern science. His renowned conflict with the Catholic Church was central to his philosophy, for Galileo was one of the first to argue that man could hope to understand how the world works, and, moreover, that we could do this by observing the real world.

Galileo had believed Copernican theory (that the planets orbited the sun) since early on, but it was only when he found the evidence needed to support the idea that he started to publicly support it. He wrote about Copernicus's theory in Italian (not the usual academic Latin), and soon his views became widely supported outside the universities. This annoyed the Aristotelian professors, who united against him seeking to persuade the Catholic Church to ban Copernicanism.

Galileo, worried by this, traveled to Rome to speak to the ecclesiastical authorities. He argued that the Bible was not intended to tell us anything about scientific theories, and that it was usual to assume that, where the Bible conflicted with common sense, it was being allegorical. But the Church was afraid of a scandal that might undermine its fight against Protestantism, and so took repressive measures. It declared Copernicanism "false and erroneous" in 1616, and commanded Galileo never again to "defend or hold" the doctrine. Galileo acquiesced.

In 1623, a longtime friend of Galileo's became the Pope. Immediately Galileo tried to get the 1616 decree revoked. He failed, but he did managed to get permission to write a book discussing both Aristotelian and Copernican theories, on two conditions: he would not take sides and would come to the conclusion that man could in any case not determine how the world worked because God could bring about the same effects in ways unimagined by man, who could not place restrictions on God's omnipotence.

The book, *Dialogue Concerning the Two Chief World System*, was completed and published in 1632, with the full backing of the censors and was immediately greeted throughout Europe as a literacy and philosophical masterpiece. Soon the Pope, realizing that people were seeing the book as a convincing argument in favor of Copernicanism, regretted having allowed its publication. The Pope argued that although the book had the official blessing of the censors, Galileo had nevertheless contravened the 1616 decree. He brought Galileo before the Inquisition, who sentenced him to house arrest for life and commanded him to publicly renounce Copernicanism. For a second time, Galileo acquiesced.

Galileo remained a faithful Catholic, but his belief in the independence of science had not been crushed. Four years before his death in 1642, while he was still under house arrest, the manuscript of his second major book was smuggled to a publisher in Holland. It was this work, referred to as *Two New Sciences*, even more than his support for Copernicus, which was to be the genesis of modern physics.

Isaac Newton [2]

The laws of motion are only part of Newton's contribution to Physical Science. He is universally recognized as one of the greatest scientists of all time, and for intellectual power his work has never been surpassed.

Newton was born in 1642 (the year in which Galileo died) in Lincolnshire. As a boy he went to King's School, Grantharn, where his name

Unit Twelve　Great Men and Women

cut with his own hands upon a window-sill, is still proudly shown today. At school he was taught Latin and grammar and showed few signs of his future genius. Indeed, he was considered dull until having been kicked by a bigger boy who was above him in class; he gave the fellow a good beating and set to work to beat him in his studies too. We are told, however, that he was very mechanically minded and fond of making windmills and model machines. This is of special interest in view of his experimental skill in later years.

At the age of nineteen he entered Trinity College Cambridge [3], where he began the study of mathematics and science, in which his great discoveries were made. In accordance with the tradition which he founded, Cambridge has maintained to the present day its position as the home of British science.

While still an undergraduate he discovered the Binomial Theorem in algebra. Just after he had taken his B. A. Degree, he did some famous experiments on the breaking up of white light into colors, and invented a new branch of mathematics known as the calculus.

At the age of twenty-six he became professor of mathematics, a post which he held until he was fifty-four. During this period his greatest discoveries were made. In 1696 he became Master of the Mint, and gave up his scientific work. He was knighted by Queen Anne in 1705. In 1727, at the age of eighty-five, he died and was buried in Westminster Abbey [4].

It was customary in Newton's time for the great mathematicians of Europe to spend months on solving a problem and then offer it as a challenge to all others. Newton always solved such problems within twenty-four hours.

He never sought fame, and many of his discoveries had to be drawn from him years after they had been made. His chief work, the *Principia* [5] (written in Latin), was published by the persuasion of his friend Halley, who paid the cost.

Many stories are told of his absent-mindedness. On one occasion a friend ate his dinner, and Newton remarked, "Dear me, I thought I had not dined, but I see I have."

On another occasion he is said to have left his guests at dinner to fetch more wine, and when after a long interval he did not return, the guests went to seek him. They found him hard at work in his study, having entirely forgotten their presence in his house.

One of his most quoted sayings is his own criticism of his discoveries: "I know not what the world may think of my labors, but to myself it seems that I have been but as a child playing on the sea-shore; sometimes finding some prettier pebble or more beautiful shell than my companions, while the unbounded ocean of truth lay undiscovered before me."

Albert Einstein[6]

Albert Einstein contributed more than any other scientist since Sir Isaac Newton to our understanding of physical reality. He was born on 14 March, 1879 in Germany. In 1901 he was granted Swiss citizenship. He avoided Swiss military service on the grounds that he had flat feet and varicose veins. Einstein worked at the patent office in Bern, Switzerland from 1902 to 1909. During this period he completed an astonishing range of theoretical physics publications, written in his spare time. The most well-known of these works is Einstein's 1905 paper proposing "the special theory of relativity". He based his new theory on the principle that the laws of physics are in the same form in any frame of reference. As a second fundamental hypothesis, Einstein assumed that the speed of light remained constant in all forms of reference.

About 1912, Einstein began a new phase of his gravitational research and called his new work the general theory of relativity. Late in 1915, Einstein published the definitive version of general theory. When British eclipse expeditions in 1919 confirmed his predictions, Einstein was idolized by the popular press.

To most people it is not easy to explain why Einstein's theory eventually shook the whole scientific and intellectual world. But because of it, scientists

Unit Twelve Great Men and Women

never again regarded the world as they had before. Basically, the theory proposed, among other things, that the maximum speed possible in the universe is that of light; that mass appears to increase with speed; and that energy and mass are equal and interchangeable. This later claim based on the formula $E=mc^2$ (Energy equals mass times the speed of light) was later proved by atomic fission, on which the atomic bomb is based. And this equation became a cornerstone in the development of nuclear energy. Later in life, When Einstein was asked to explain his law of relativity to a group of young students, he said: "When you sit with a nice girl for two hours, you think it's only a minute. But when you sit on a hot stove for a minute, you think it's two hours. That is relativity."

Einstein received the Nobel Prize in 1921 not for relativity but for his 1905 work on the photoelectric effect. A third visit to the United States in 1932 was followed by the offer of a post at Princeton. Einstein accepted and left Germany in December 1932 for the United States. The following month the Nazis came to power in Germany and Einstein was never to return there.

At Princeton his work attempted to unify the laws of physics. In 1940 Einstein became a citizen of the United States, but chose to retain his Swiss citizenship.

He made many contributions to peace during his life. In 1939 Einstein collaborated with several other physicists in writing a letter to President Franklin Roosevelt, pointing out the possibility of making an atomic bomb and the likelihood that the German government was embarking on such course. The letter helped lend urgency to efforts in the U. S. to build the atomic bomb. In 1944 he made a contribution to the war effort by handwriting his 1905 paper on special relativity and putting it up for auction. It raised six million dollars, the manuscript today being in the Library of Congress. After the war, Einstein was active in the cause of international disarmament and world government.

One week before his death, Einstein signed his last letter. It was a letter to

Bertrand Russel in which he agreed that his name should go on a manifesto urging all nations to give up nuclear weapons. He died in Princeton, New Jersey, U. S. A. on 18 April, 1955.

New Words and Expressions

Copernican 哥白尼学说的
Aristotelian 亚里士多德的
ecclesiastical （基督）教会的；教士的
allegorical 比喻的，寓言的
scandal 反感；愤慨；物议；流言蜚语
undermine 暗中破坏；逐渐损害（或削弱）
Protestantism 基督教新派；耶稣教
erroneous 错误的
acquiesce 默认，默许；勉强同意
Pope 罗马天主教教皇
decree 教令
revoke 撤销；解除
omnipotence 无限权力，无限权威
censor （书刊、报纸、电影等的）审查员，审查官
blessing 同意，允准
contravene 违反
Inquisition 中世纪天主教审判异端的宗教法庭；宗教裁判所
manuscript 手稿
smuggle 偷运，偷带
genesis 创始，起源
surpass 超越，胜过
window-sill 窗沿，窗台
mint 铸币厂
knight 封……为爵士
customary 习惯的，按照惯例的
pebble 小圆石，小鹅卵石
unbounded 无边际的；无限的
gravitational 万有引力的，重力的
idolize 将……当作偶像崇拜
cornerstone 基石，柱石；地基
Nazis 纳粹
collaborate 合作，协作
embark (on) 发动；开始工作
auction 拍卖
disarmament 解除武装，缴械
manifesto 宣言，告示

Notes

1. Galileo Galilei（1564—1642）伽利略，意大利物理学家、天文学家、哲学家和近代实验科学的先驱。

Unit Twelve　Great Men and Women

2. Isaac Newton（1643—1727）艾萨克·牛顿，物理学家、数学家和哲学家。

3. Trinity College Cambridge 剑桥大学三一学院。剑桥大学是一所综合性大学，三一学院是剑桥大学最大的一个学院。Trinity 意为"三位一体，三个一组的（物）"，教会中的三位一体指的是"圣父""圣子"和"圣灵"。

4. Westminster Abbey 威斯敏斯特教堂是英国著名的基督教堂。1050 年由英王爱德华开始兴建，后又重建多次。该教堂是英国国王加冕和历代国王及著名人物埋葬之地。

5. Principia《自然哲学的数学原理》。1687 年由哈雷资助出版。该书将自然界的各种现象用数学的规律加以说明。书中的自然哲学就是自然科学的意思。

6. Albert Einstein（1879—1955）阿尔伯特·爱因斯坦，美籍德国犹太裔、理论物理学家、相对论的创立者和现代物理学奠基人。

Exercises

I. Find out the English equivalents to the following Chinese terms from the passage.

1. 天主教会
2. 镇压措施
3. 被软禁
4. 二项式定理
5. 造币厂厂长
6. （永久性）静脉肿大或曲张
7. 参照系
8. 基本假设
9. 日食测试
10. 各种大众传媒
11. 整个科学和知识界
12. 原子裂变
13. 光电效应
14. 国际裁军事业
15. 核武器

II. Choose the best answer according to the information in the passage.

1. Which of the following statements is true?

   A) Einstein would have joined the army if he hadn't had flat feet and varicose veins.

   B) Einstein began his research work after he was granted Swiss citizenship.

   C) Einstein contributed more to science than any other scientist in history.

   D) Einstein put forward the special theory of relativity in a paper written in his spare time.

2. Einstein received the Nobel Prize _____.

   A) because he proposed the Theory of Relativity

   B) because the equation $E=mc^2$ became a cornerstone in the development of nuclear energy

   C) for his work on the photoelectric effect

   D) for his work in world peace

3. He left Germany for America because _____.

   A) he received the Nobel Prize

   B) he accepted a post in America

   C) the Nazis came to power

   D) he loved peace

4. In Einstein's opinion _____.

   A) atomic bombs can help America become a superpower

   B) with an atomic bomb America is able to bully any other country in the world

   C) the speed of light remains constant in frames of reference

   D) the laws of physics can be changed at will

5. Which of the following statements about Einstein is NOT true?

   A) His Theory of Relativity had a profound and startling influence on science and culture.

Unit Twelve   Great Men and Women

B) His famous equation led to the development of atomic energy and weapons.

C) He was active in the cause of world peace.

D) His excellent work at the theory of relativity won him the Nobel Prize.

Part C Extended Reading

Madame Curie and Radium

Madame Curie will always be remembered as the discoverer of an element, radium, which had never been found before her day.

A certain scientist had discovered that a metal called uranium gave off a kind of radiation, which Marie Curie [1] was later to call radioactivity. But where did this radiation come from, and what was it like? Here was a secret of nature which she set out to discover. Only a scientist could understand all that this pursuit meant. The experiments were done most carefully again and again. There was failure, success, more failure, a little success, a little more success. All seemed to prove that in the mineral which she was examining there was some form of radiation which man knew nothing about.

Four years before this, Marie had expressed her thoughts in words much like these: "Life is not easy for any of us. We must work, and above all we must believe in ourselves. We must believe in ourselves. We must believe that each one of us is able to do something well, and that, when we discover what this something is, we must work until we succeed." This something in Madame Curie's own life was to lead science down a new path to a great discovery.

At this time her husband left his own laboratory work, in which he had been very successful, and joined with her in her research for this unknown radiation. In 1898 they declared that they believed there was something in nature which gave out radioactivity. To this something, still unseen, they gave the name radium. All this was very interesting, but it was against the beliefs of some of the scientists of that day. These scientists were very polite to the two Curies, but they could not believe them. The common feeling among them was: "Show us some radium, and we will believe you."

There was an old building at the back of the school where Pierre Curie[2] had been working. Its walls and roof were made of wood and glass. It was furnished with some old tables, a blackboard, and an old stove. It was not much better than a shed, and no one else seemed to want it. The Curies moved in, and set up their laboratory and workshops. Here for four very difficult years they worked, every moment that they could spare, weighing and boiling and measuring and calculating and thinking. They believed that radium was hidden somewhere in the mass of mineral dirt which was sent to them from far away. But where?

The shed was hot in summer and cold in winter, and when it rained, water dropped from the ceiling. But in spite of all the discomforts, the Curies worked on. For them these were the four happiest years of their lives.

Then, one evening in 1902, as husband and wife sat together in their home, Marie Curie said: "Let's go down there for a moment." It was nine o'clock and they had been "down there" only two hours before. But they put on their coats and were soon walking along the street to the shed. Pierre turned the key in the lock and opened the door. "Don't light the lamps," said Marie, and they stood there in the darkness. "Look!...Look!"

And there, glowing with faint blue light in the glass test-tubes on the tables, was the mysterious something which they had worked so hard to find: Radium.

New Words and Expressions

radium 镭
uranium 铀
furnish 为（房间等）配备家具
shed 工作棚，棚屋，小屋

Notes

1. Marie Curie（1867—1934）玛丽·居里，著名的女性物理学家，两度获得诺贝尔奖，与其丈夫共同发现了放射性元素镭。

2. Pierre Curie（1859—1906）皮埃尔·居里，法国著名的物理学家，居里夫人的丈夫，"居里定律"的发现者。

Exercises

Answer the following questions by using the sentences in the text or in your own words.

1. What does the "secret" mean in Paragraph 1?
2. Marie said: "Life is not easy for any of us. We must work, and above all we must believe in ourselves. We must believe in ourselves. We must believe that each one of us is able to do something well, and that, when we discover what this something is, we must work until we succeed." In her words, what does "this something" refer to?
3. What was the common feeling toward Curie couple's finding among these scientists?
4. Where did the Curies set up their laboratory and how did they work?
5. When did the Curies see the real radium?

References

陈羽伦. 1995. 科技英语选粹. 北京：中国对外翻译出版公司.

崔秀敏. 2002. 英语实用能力辅导教程. 北京：外语教学与研究出版社.

戴丹青，胡晓军. 2001. 读点科学史. 上海：上海科技教育出版社.

樊振帼. 1993. 江苏省职称外语考试辅导（阅读理解与翻译分册）. 南京：译林出版社.

高凤平. 2006. CET-6 科普阅读. 北京：世界图书出版公司.

G. C. 索恩利编著. 1981. 英语科普文选（第三集）. 京广英译. 北京：科学普及出版社.

郭富强. 2004. 英汉翻译理论与实践. 北京：机械工业出版社.

郭继荣，赵冬梅，史焱. 2000. 英语科普阅读. 西安：西安交通大学出版社.

郭庆民. 2000. 英语复习指南（上册）阅读理解. 北京：世界图书出版公司.

胡春洞，王才仁，程世禄，张国扬. 1999. ESP 的理论与实践. 南宁：广西教育出版社.

黄忠廉，李亚舒. 2004. 科学翻译学. 北京：中国对外翻译出版公司.

姜维焕. 2000. 英语科普阅读. 西安：西安交通大学出版社.

金焕荣. 2002. 科技英语阅读. 苏州：苏州大学出版社.

金泉元，张长缨. 1994. CET-6 成功之路：大学英语六级考试题透视. 北京：宇航出版社.

李庆明，陈国新，莫文莉. 2000. 英语科普阅读. 西安：西安交通大学出版社.

李维屏，杨理达. 1989. 实用英语写作. 上海：华东理工大学出版社.

李运兴. 2003. 英汉语篇翻译（第二版）. 北京：清华大学出版社.

林钜洸. 2004. 当代高科技及其哲理. 北京：外语教学与研究出版社.

刘可友，廖译.1996.大学英语阅读能力训练16法.武汉：武汉测绘科技大学出版社.

潘能.2002.英语阅读八步.西安：西安交通大学出版社.

潘永樑.2002.新世纪英语科普阅读.上海：上海外语教育出版社.

秦秀白.2002.英语语体和文体要略.上海：上海外语教育出版社.

司树森，刘典忠.2005.Super·分级分类英汉对照读物·四级：科普篇.北京：中国人民大学出版社.

孙湘生.1993.突破大学英语四级阅读难关（上、下册）.长沙：湖南师范大学出版社.

王令坤，朱俊松，朱慧敏，李振国，葛纪红.2002.科技英语读写译教程.北京：外语教学与研究出版社.

王守仁.2016.《大学英语教学指南》要点解读.外语界，(3)，5-13.

王卫平，潘丽蓉.2009.英语科技文献的语言特点与翻译.上海：上海交通大学出版社.

王忠樑.1999.怎样学好考好大学英语——阅读篇.上海：上海世界图书出版公司.

王佐良，丁往道.1987.英语文体学引论.北京：外语教学与研究出版社.

翁凤翔.2002.实用翻译.杭州：浙江大学出版社.

吴玮翔，石高玉，彭友良.1990.CET, EPT, TOEFL阅读理解应试技巧.南京：河海大学出版社.

谢小苑.2008.科技英语翻译技巧与实践.北京：国防工业出版社.

谢小苑.2014.科技英语阅读.北京：国防工业出版社.

谢小苑.2015.科技英语翻译.北京：国防工业出版社.

徐钟.2001.新大纲大学英语：阅读技能精编.上海：上海世界图书出版公司.

许建平.2003.英汉互译实践与技巧.北京：清华大学出版社.

严俊仁.2000.科技英语翻译技巧.北京：国防工业出版社.

俞宝发.2002.英语背诵范文精典.上海：上海三联书店.

翟天利.2003.科技英语阅读与翻译实用教程.北京：新时代出版社.

张玲. 1999. 英语阅读 10 技能. 北京：宇航出版社.

张亚非. 2002. 现代科技英语教程. 北京：科学出版社.

朱万忠. 2000. 英语阅读技能讲·练·测. 重庆：重庆大学出版社.

朱永生，郑立信，苗兴伟. 2001. 英汉语篇衔接手段对比研究. 上海：上海外语教育出版社.

Carlisle, R. 2004. Scientific American Inventions and Discoveries. New Jersey: John Wiley & Sons.

McClellan, J. E., & Dorn, H. 2006. Science and Technology in World History: An Introduction (Second Edition). Maryland: The John Hopkins University Press.